# REBEL CELL

## CANCER, EVOLUTION, AND THE NEW SCIENCE OF LIFE'S OLDEST BETRAYAL

KAT ARNEY

BenBella Books, Inc.
Dallas, Texas

This book is for informational purposes only. It is not intended to serve as a substitute for professional medical advice. The author and publisher specifically disclaim any and all liability arising directly or indirectly from the use of any information contained in this book. A health care professional should be consulted regarding your specific medical situation. Any product mentioned in this book does not imply endorsement of that product by the author or publisher.

BenBella Books, Inc.
10440 N. Central Expressway, Suite 800
Dallas, TX 75231
www.benbellabooks.com
Send feedback to feedback@benbellabooks.com



Printed in the United States of America
10 9 8 7 6 5 4 3 2 1

Library of Congress Cataloging-in-Publication Data is available upon request.
LCCN: 2020941481
ISBN 9781950665303 (trade paper)
ISBN 9781950665518 (ebook)

Originally published by The Orion Publishing Group, Ltd.
Cover illustration by Sarah Avinger and texture by Lost & Taken
Typeset by Input Data Services Ltd, Somerset
Printed by Lake Book Manufacturing

Distributed to the trade by Two Rivers Distribution, an Ingram brand
www.tworiversdistribution.com

# PRAISE FOR *REBEL CELL*

"This book is packed with big ideas about life. Every chapter has something in it which made me think *wow*. Having worked in a major cancer charity for many years, Arney writes with genuine in-depth understanding and is a perfect guide."

**—Daniel M. Davis, author of *The Beautiful Cure***

"Better than just a history or scientific rundown of cancer, Kat Arney presents a philosophy for how to think about cancer."

**—Zach Weinersmith, coauthor of *Soonish* and creator of *Saturday Morning Breakfast Cereal***

"Kat Arney is a science writer for all of us—a powerful and talented storyteller."

**—Stephen McGann, author and actor on *Call the Midwife***

"A crystal clear reappraisal of the story behind that word we fear to mention. If you want to know your enemy, read this book. A myth-busting masterclass in science writing."

**—Dallas Campbell, science broadcaster and author of *Ad Astra***

"Forget magic bullets and much-hyped miracle cures; to improve the chances of cancer patients we need revolutionary new thinking. And that new thinking, Arney forcefully argues, is evolutionary... This lively, scholarly, and accessible book should be essential reading for oncologists the world over and powerfully demonstrates that nothing in biology, cancer included, makes sense except in the light of evolution."

**—Laurence D. Hurst, director of the Milner Centre for Evolution and president of the Genetics Society**

"Kat Arney does it again: taking the complex and making it easy to grasp, demystifying the mysterious, asking the right questions and finding the surprising answers—and all with rollicking good humor and bonhomie. The war on cancer would itself be far more advanced if its practitioners knew how to communicate with us and each other as well as Kat does in this mind-thrilling page turner."

**—Mark Stevenson, futurist and author of *We Do Things Differently: The Outsiders Rebooting Our World***

# CONTENTS

*To life, love, and loss*

*Turning and turning in the widening gyre*
*The falcon cannot hear the falconer;*
*Things fall apart; the center cannot hold;*
*Mere anarchy is loosed upon the world*

W.B. Yeats

# INTRODUCTION

"Cancer starts when a cell picks up genetic mutations and multiplies out of control."

I don't know how many times I've written some variation of that sentence during my career as a science writer, including the twelve years I spent on the communications team at the world's leading cancer research charity. Never once did I stop to consider what it really means. Or that it might be wrong.

Cancer is a disease that affects us all. Even if you've been lucky enough not to see its effects up close, either in yourself or in someone you love, cancer is a global health problem that kills millions of people around the world every year. Scientists and doctors have been trying to discover its causes, consequences, and cures for thousands of years, arguably only making any significant progress in the latter half of the twentieth century. Today, around half of the people diagnosed with cancer in the UK can expect to survive for ten years or more—a figure that's only likely to rise in the future. To an optimist, this is a glass half full.

We already know how to cure cancer. Or rather, we already know how to cure *some* cancers. The best way is to spot them as soon as possible and remove them with careful surgery before they've started spreading round the body (known as metastasis). Radiotherapy can be curative and hormone therapy can be very effective for keeping breast and prostate cancers at bay, if used at the right time. Many blood cancers respond strikingly well to chemotherapy—particularly in children—and drugs can completely cure testicular cancer even at an advanced stage. There have been startling results with the new generation of immunotherapies, yet they currently only work for fewer than one in five patients who take them. But for most of the unlucky ones whose disease has started its unrelenting march through the body, the question changes from "Will I get better?" to "How long have I got left?" Not *if*, but *when*.

This is pretty much the same situation we've been in since President Richard Nixon infamously declared "War on Cancer" in 1971. Seeking a distraction from the conflict in Vietnam and hoping to capitalize on the pioneering spirit engendered by the recent Apollo moon landings, Nixon pledged millions of dollars toward the quest for a cure within a decade. Alas, in an unfortunate parallel with the situation in the Far East, he had wildly underestimated his enemy. In 1986, statistician John Bailar ran the numbers: despite a few successes here and there, the vast majority of late-stage cancers still remained stubbornly incurable. In Bailar's own words, the War on Cancer was to be judged a "qualified failure."

Although there's been a fair bit of successful tinkering around the edges for certain types of cancer—most notably malignant melanoma—a close look at today's statistics reveals the same patterns. More and more people are being diagnosed at an early stage when treatment is much more likely to be effective, boosting the overall figures. But survival from advanced metastatic cancer is still likely to be measured in months or single-digit years rather than decades.

The big problem is that precision tools of surgery and radiotherapy are virtually useless against rampant disease, and chemotherapy is a blunt weapon based on the principle of killing cancer cells faster than it kills the healthy ones. Even when it works, the tumors almost inevitably come back—weeks, months, or even years later—and every subsequent round of treatment is a more brutal assault on health, with diminishing returns. The empty half of the glass is proving difficult to fill.

At the turn of the twentieth century, scientists working for the newly founded Imperial Cancer Research Fund in London were busying themselves growing cancerous mouse cells in the laboratory in the hope of figuring out the secret of their prodigious multiplication. The researchers marveled at the seemingly inexhaustible regenerative capacity of the cells in their care, with General Superintendent of Research Ernest Bashford noting in the charity's 1905 annual scientific report that "under artificial propagation a mouse tumor has produced an amount of tissue sufficient to yield a giant mouse as large as a St. Bernard."

Today, we have a more complete picture of what happens when cells throw off their molecular shackles. Cheats emerge from the civilized multicellular society of cells, growing and dividing out of control in a shambolic mockery of normal life. One cell becomes two, two cells become four, four become eight, piling up to make a multimillion-strong mob. But they don't stop there. These rebels invade and corrupt the well-behaved tissues around them, persuading the immune system—the body's police force—to look the other way. They slip unnoticed into the bloodstream, traveling through arteries and veins to set up splinter groups and sleeper cells. Every single one of them is driven by rogue versions of our own genes—the genetic instruction manual that tells cells when to divide, what to become, and even when to die.

There's a long-standing belief that the "cure for cancer" lies in understanding the faulty genes and molecules within tumor cells—a task that has occupied a small army of scientists for the best part of a century, at a cost of countless billions of dollars. Researchers have extracted, read, and analyzed DNA from tumors and healthy tissue samples from thousands of cancer patients around the world: endless letters spelling out the recipe book of life and the typos within it that are thought to be responsible for driving cancers to grow and spread. But rather than providing clarity, this information reveals more than ever before about the genetic chaos within tumors.

We can see the scars left in the genome by tobacco smoke or ultraviolet light from the sun. There's evidence showing how the biological defense mechanisms that are meant to protect our cells can fail, or even turn against us. There are strange marks with unknown causes, which may one day be pinned down to damaging chemicals in the environment or new molecular processes. DNA analysis has revealed remnants of large- and small-scale damage, from a handful of typos to scenes of epic genetic disaster as whole chromosomes are shattered and stitched back together again. To make things even more confusing, it now turns out that even perfectly healthy tissues are a mess of mutated cells by the time we get to middle age, many of them carrying what would normally be classified as cancerous mutations.

More disturbingly, these studies have shown that the genetic changes that turn a single cell into a tumor aren't consistent or fixed. There is no one "cancer gene" just as there is no one "cancer cure." There are major differences in the genetic makeup of tumors from person to person, and even variations in gene faults across the miniature landscape of an individual tumor. Every cancer is a genetic patchwork made up of distinct groups of cells, any one of which could be carrying gene changes that make it immune to the effects of treatment. Once a cancer has grown to a certain size and diversity, relapse is inevitable.

Scientists have begun to see the progress of cancer as a microcosm of evolution, with cells picking up new mutations and undergoing natural selection as they grow and spread, akin to Darwin's great tree of life. And it's here we discover another uncomfortable biological truth about cancer: the very processes that have driven the evolution of life on this planet are inescapably at work within our own bodies as cancer develops.

To make things worse, some of these selective pressures come in the form of supposedly life-saving therapies, which strip away drug-sensitive cancer cells and allow resistant ones to flourish. Unfortunately, what doesn't kill cancer only makes it stronger and when it comes back, it's unstoppable. It's no wonder that our current approaches to treatment are powerless against such a malignant monster.

We urgently need a new way of thinking about how cancer arises, and how we might prevent and treat it based on this evolutionary reality. We need a clearer understanding of the species of rogue cells evolving within a tumor and the landscape in which they live, seeing them as populations changing over time, not fixed entities that can be described with a simple list of mutations. German biologist Richard Goldschmidt coined the term "hopeful monsters" to describe organisms that had evolved dramatic new properties in a relatively short time frame in the prehistoric Cambrian seas. Cancer cells are "selfish monsters," evolving wildly and rapidly in the space of a patient's lifetime. And just as famine or predators act as important selective pressures that shape species, cancer cells respond to selection too, acting out an evolutionary drama within the ecosystem of the body.

In this brave new world, where every cancer is genetically unique and evolves its way out of trouble, the old models of drug development and clinical trials no longer stand up. It's become a hopelessly bureaucratic business, using ever more sophisticated tools to achieve a shrinking return. We need to be a lot smarter to beat such a wily foe. But we're finally starting to decipher cancer's secret evolutionary playbook, as well as revealing the ecology of the landscape in which these rogue cells live. And there is growing hope that we can use this knowledge to predict and confound its next moves, skillfully manipulating the processes of evolution itself to steer and shape exuberant tumor growth.

In January 2019, as I was working on the first draft of this book, my Twitter feed lit up with news that an Israeli biotech firm had apparently developed a cure for all cancers that would be available within a year. Despite the credulous retweets and media reporting, the therapy had only been tested in mice and had no clinical data to support these claims, suggesting that the announcement was likely to do more for the health of the company's finances rather than any cancer patient in the foreseeable future. Somewhat predictably, a year later this "miracle cure" was still in development and not a single patient had been treated.

Infuriatingly, articles debunking such overhyped miracle cures and outright nonsense usually receive many times fewer clicks than the original coverage. This is hardly a modern problem. In 1904, Sir D'Arcy Power, a surgeon at St. Bartholomew's Hospital in London, wrote a furious paper in the *British Medical Journal* railing against a quack cancer cure produced by one Dr. Otto Schmidt, a German doctor. He noted that Schmidt's ineffectual treatment "gained a somewhat wider circulation than was intended, as a long abstract appeared in the *Daily Mail*."

We want to believe that there is a Cure for Cancer—a term that has embedded itself deep in our cultural consciousness to mean total eradication of the disease. We want to know that all the time, money, effort, pain, and lost lives are getting us closer to finding such a cure. We are easily seduced by talk of smart drugs, magic bullets, and miracles. Shifting toward a new way of evolutionary and ecological thinking about cancer is going to require a change in mindset—not just from the scientific and medical community, but from patients and the public—if the long-awaited solution doesn't look quite like what we expected.

This isn't a story about cancer. It's a story about life. I want to show you that cancer isn't a modern human disease, but is hardwired into the fundamental processes of biology. We'll discover how the roots of this rebellion go all the way back to the origins of multicellular life, setting up the organized structures from which cheating cells emerge. Looking back over more than a century of research, we'll see how scientists have learned the genetic secrets of cancer—knowledge that has been both revolutionary and misleading at the same time. We'll find out how the same evolutionary forces that shape the spectacular diversity of life on Earth also act down at the level of rogue cells and how we must learn to work with them rather than against them in order to beat cancer. And while we can't deny our own biology—and nobody lives for ever—we'll look forward to a future when every person who is told, "You have cancer" then hears the words, "It's OK—we know what to do about it."

# 1

# LET'S BEGIN AT THE VERY BEGINNING

It all starts with one.

Though there may have been many other similar objects floating around in the primordial soup roughly 3.8 billion years ago, LUCA was the cell that got lucky.* Spawned in the hot, dark, and suffocating environment around ancient deep-sea hydrothermal vents, LUCA was a simple bacteria-like cell that had somehow managed to accumulate all the components necessary for independent life: a set of molecular machinery and genetic instructions enabling it to generate energy, maintain itself, and—most importantly—replicate.

One cell became two. Two became four. Four became eight, and on and on and on and on. And now, billions

* LUCA—the Last Universal Common Ancestor—is the name given to the most recent organism from which all life on Earth evolved [although nearly 4 billion years is arguably stretching the concept of "recent" to breaking point].

of years later, here we are. Every single cell in your body, every cell in the tree outside your window, every cell in the goldfinch that chirps in its branches or in the colonies of bacteria lurking in your toilet bowl can trace its origins all the way back to LUCA through an unbroken chain of cell divisions. This process of cell replication is the fundamental engine that drives the profusion of life on Earth. It's what turns an acorn into an oak, a lump of yeasty dough into a loaf of fluffy bread, a fertilized egg into a baby, and a cancer cell into a deadly tumor.

## ANCIENT AND MODERN

When someone hears the news that they have cancer, one of the first questions they often ask is, "Why me?" But the first question I want to ask is, "Why us?"

If you follow the headlines declaring ever-rising cancer rates, it's easy to fall into the trap of thinking that cancer is a recent disease caused by our unhealthy modern lifestyles. But given that cancer will inevitably emerge in almost any multicellular species, this simply isn't true.

In October 2010, while I was working in the science communications team at the charity Cancer Research UK, the University of Manchester put out a press release about a review written by two researchers, Rosalie David and Michael Zimmerman, published in the journal *Nature Reviews Cancer*. They concluded that because cancer is rare in Egyptian mummies and other ancient remains, it must be an almost entirely modern confection for which we only have ourselves to blame. Unsurprisingly, the story was media catnip. It quickly appeared in newspapers and online, stirring me into action to write a post for the charity's blog arguing that these claims were not only misleading, but also wrong.

For a start, rare doesn't mean non-existent. We have no way of knowing whether or not the proportion of cancers that turn up in the archaeological records is an accurate reflection of the health of the populace in which they arose. Calculating accurate cancer incidence statistics for long-dead human populations is almost impossible, given the relatively small proportion of remains that have been unearthed from ancient times in comparison to the number of humans who have ever lived. What's more, cancer is a disease that mostly affects older people, with the incidence rising sharply over the age of sixty. Many modern populations are lucky enough to avoid the hazards that sent our ancestors to early graves, such as infectious diseases, poor diets, death in childbirth, and generally unpleasant living conditions. But as average life expectancy has significantly increased around the world, so too have the chances of living to an age where the risk of cancer becomes an issue.

You might make it to fifty or more in ancient Egypt if you were wealthy and well fed, but the poor folk would be lucky to scrape past thirty. In fifteenth-century England, men could expect to live to fifty on average, with women only reaching around thirty, presumably thanks to high rates of death in childbirth. While archaeologists can have a good stab at guessing the age of an individual they've unearthed, perhaps by looking at the condition of their teeth and bones or any accompanying artifacts, it's very difficult to work out an age-standardized cancer incidence curve for people who shuffled off this mortal coil many thousands of years ago.

Secondly, most archaeological specimens are little more than skeletons. Although some cancers leave their mark in the bone, others are more likely to stay confined to swiftly decomposing internal organs. The fact that tumors have been spotted in a number of mummified bodies, whose soft tissues are preserved, doesn't particularly scream "extremely rare" to me. Certainly, the disease was common enough to be worth mentioning by ancient Egyptian, Roman, and Greek doctors, with the second-century Greek physician Galen noting, "We have often seen in the breasts a tumor . . . This disease we have cured often in its beginning, but when it has progressed to a substantial size no one can cure it without surgery." As we'll see, more than 275 examples of cancer have been documented in people who lived before the turn of the twentieth century, including extremely rare childhood tumors as well as more common cancers. And those are just the ones we know about. How many of Galen's breast cancer patients are lost to history because we have no physical or written trace of their existence?

In fact, the original review is far more circumspect than the press stories. Michael Zimmerman is a respected scientist who has produced detailed studies of tumors in mummies, and the paper goes into great detail about the archaeological and cultural evidence for cancer in antiquity. We could argue endlessly about whether or not this meets a definition of "rare," but by far my biggest issue with the story lay in the university's original press release, which contains the following quote from Rosalie David: "There is nothing in the natural environment that can cause cancer. So it has to be a man-made disease, down to pollution and changes to our diet and lifestyle."

Sorry, but no. It simply isn't true that the ancient past was some kind of wellness utopia. As we'll see in the next few chapters, although modern lifestyles and habits can undoubtedly increase the risk of cancer, the natural environment is awash with things that cause cancer, from viruses and other infectious diseases to food molds and naturally occurring chemicals in plants (even the "organic" ones). Radioactive radon gas seeps out of the ground in many parts of the world as a result of natural processes, especially areas rich in volcanic rocks. It's thought to be responsible for the unusually high rates of cancer discovered in the remains of a group of villagers living in the American Southwest about a thousand years ago. Our very own sun showers us with cancer-causing ultraviolet radiation every day. Carcinogenic compounds abound in the soot and smoke produced by open fires, used by humans for food and warmth for more than a hundred thousand years and particularly noxious in confined spaces like caves or kitchens. And most childhood cancers have very little to do with any environmental factors, instead arising as a consequence of normal developmental processes running amok (see page 138).

To get a better idea of the way that cancer has haunted the human race throughout our history, I met up with Casey Kirkpatrick, one of the co-founders of the Paleo-oncology

Research Organization (PRO)—a small but determined group of women scientists dedicated to investigating cancer throughout antiquity. They're following in the footsteps of a handful of pioneers of ancient disease research (paleopathology), notably doctor-turned-Egyptologist Eugen Strouhal and American anthropologist Jane Buikstra, and taking a very systematic approach to the problem. One of the PRO's first projects was to set up the Cancer Research in Ancient Bodies Database (known as CRAB, as a nod to the ancient etymology of the disease, page 216), where they gathered all the information they could find about cancer in humans who lived before the twentieth century.

It's still a work in progress, but at the time of writing there were around 275 entries—significantly more than had been described by 2010 when Zimmerman and David's review came out. That still might not seem like a lot, but there will be many more ancient cancers out there that have simply gone unnoticed. It is, after all, remarkably difficult to diagnose someone who's been dead for more than a thousand years, especially if all you have to go on are a few bits of bone.

The main tools for diagnosing cancer in ancient remains are X-rays and CT scans. In fact, the first X-ray image of a mummy was published by pioneering English Egyptologist Flinders Petrie in early 1896, just four months after the discovery of X-rays (although he was searching for jewels or amulets hidden under funereal wrappings rather than tumors). The first cancers in mummies were spotted in the 1950s, but the development of three-dimensional CT scanning in the

1970s was a game changer. Archaeologists could now virtually unwrap mummies and see what lay within, leading to the identification of many more cases.

Spotting a strange lump or abnormal structure in an ancient skeleton or mummy doesn't automatically signify cancer: it could be a benign tumor, a cyst, or any one of a number of other diseases. It might be a sign of fluorosis—a condition in which soft tissue turns into bone owing to high levels of fluorine in the environment, commonly found near volcanoes. Or it could be something known as pseudopathology, where the normal decomposition of bone creates an illusion of disease. However, there are some giveaway clues.

Certain cancers look very distinctive—what Casey Kirkpatrick and her colleagues would describe as having a specific *pathognomic*. Others aren't so obvious. While CT scanning and X-rays might reveal there's some kind of cancer present, it can be hard to say exactly which type, so the best a paleopathologist can offer is a selection of options rather than a definitive answer. Myeloma—a cancer affecting white blood cells in the bone marrow—leaves the same kind of traces in bones as tumors that have spread through the body from elsewhere, while the blood cancers leukemia and lymphoma are virtually indistinguishable in ancient remains. And while a modern-day patient with suspected cancer will be put through a systematic battery of tests and scans to pin down their disease, there isn't a similar standardized pathway for assessing cancer in ancient remains—something the PRO team is working to address.

Another problem is figuring out how diseases manifested

within in the body in the distant past. There are big differences around the world in the causes, numbers, and types of cancer that turn up in modern populations, and it's rare that someone living in a wealthy country today would die of cancer without receiving any treatment at all. So trying to compare 4,000-year-old Egyptians, third-century Inuit, or precolonial Peruvian villagers with modern Westerners is a tricky task. Some researchers are trying to make more realistic comparisons with less developed cultures and populations without good access to medical care, although gathering accurate data and statistics in these parts of the world can be challenging.

The difficulty of diagnosis has led to long-running arguments about whether or not strange lumps and bumps found in ancient remains are true examples of cancer or if they could have been caused by other means. One of the most famous (and controversial) examples is the mass bulging from the jawbone of Kanam Man, a fossilized remnant of an ancient human dug up by fossil-hunter Louis Leakey and his team near the Kenyan shore of Lake Victoria in 1932. The exact age of the fossil and its position in our ancestral family tree is disputed—although it's thought to be at least 700,000 years old—as is the true nature of the mass protruding from its surface. If it is, as some people argue, the remnants of a bone tumor or Burkitt's lymphoma, then this lump is one of the oldest hominin cancers that we know of. Or, as others believe, it could just be overgrown bone resulting from a poorly healed fractured jaw.

Other contentious examples are an apparent spinal tumor

in the fossilized skeleton of a young australopithecine—one of our ancient primate ancestors in east Africa, who lived nearly 2 million years ago—and a strange growth in a rib belonging to a 120,000-year-old Neanderthal from what is now Krapina in Croatia. That last one is most likely the result of a non-cancerous condition called fibrous dysplasia, where normal bone is gradually replaced with weak fibrous tissue.

A more definite diagnosis comes from a toe bone from the South African Swartkrans cave—the "Cradle of Humankind" where our species is first thought to have emerged. Although it's impossible to pin a precise species on the bone, which dates back more than 1.6 million years, it very likely belonged to an individual who was related to humans. Unfortunately, they were probably afflicted with a type of aggressive bone cancer known as osteosarcoma, which tends to affect teenagers and isn't known to be related to anything in the environment or lifestyle. This is the oldest known identifiable cancer in a human ancestor to date, but that may change in the future as more bones are discovered and diagnostic techniques improve.

There are many other examples of possible ancient cancers dotted around the world. A benign tumor was spotted in the 250,000-year-old jawbone of an adult *Homo naledi*—a member of the most recently identified group of extinct human ancestors, whose numerous bones were discovered in the Rising Star cave system in South Africa in 2015. There's a skull bone belonging to an ancestor of the Neanderthals, *Homo heidelbergensis,* who possibly died of a brain tumor up to 350,000 years ago in the region of Europe we now call Germany. Then there's Lemdubu woman—a sturdy, rugged-jawed, twenty-something female who was buried 18,000 years ago in an Indonesian cave. Her bones are peppered with holes that look just like the cavities caused by metastatic cancer. Frustratingly, ancient fossilized skeletons don't come with neatly preserved medical notes, so we may never know the truth about these long-gone souls.

New molecular biology techniques offer a potential way forward. As DNA detection techniques have become more sensitive and much cheaper, researchers can now analyze tiny fragments of DNA gathered from historic remains. This tactic has most famously been used in the case of the mummified body of Italian Renaissance ruler King Ferrante I of Aragon, which was found to harbor an exceptionally well-preserved tumor in the pelvis. Under the microscope, the cancer cells looked like they could have either started life in the king's bowel or in his prostate. Genetic testing revealed that the tumor carried a fault in a gene called KRAS, which is common in bowel tumors but virtually unknown in prostate cancer. This provided Ferrante with a definitive diagnosis a mere 500 years after his death.

However, genetic techniques have limited usefulness, because they rely on capturing a sample of DNA from a tumor, either in a preserved organ or where it's spread into the bone. And it may be of limited benefit now that we know even normal cells can contain apparently "cancerous" mutations (see page 107). An alternative idea is to look for faulty protein molecules that might be a more reliable indicator of cancer—an approach known as proteomics. It's more technically challenging and expensive to identify proteins than it is to do more straightforward DNA sequencing, so proteomic analysis tends to be reserved for the most special samples in a paleopathologist's collection. Costs are coming down all the time, though, so it's likely to be more widely applied in the future.

Despite the increasing availability of tools, the limiting factor will always be the supply of human remains on which to use them. Perfectly statistically balanced populations of skeletons can't be magicked out of the ground—you get what you get and you get on with it. There's also something called the "osteological paradox," first proposed in 1992 by anthropologist James Wood and his colleagues, which says that any archaeological record will never be truly representative of the state of pathology in that population. This is partly because some people will succumb quickly to diseases that leave no trace in their mortal remains and also because you can only ever know about someone's health at the point of their death. For example, finding the skeleton of a fifteen-year-old girl who died 2,000 years ago tells you nothing about the health of her friends who survived to an older age. But we do know that many different types of cancer have been found all over the world in many cultures spanning many thousands of years, including what would be considered as very rare tumor types by today's standards.

There are other, more nebulous things that affect whether researchers are more or less likely to find certain types of people and diseases in the archaeological record or information about them. If someone had a very fast-growing cancer, they might just suddenly die without it ever being diagnosed or leaving a mark on their bones. Even if an autopsy was performed, many cultures have a stigma around cancer, believing it's sinful or infectious, so families might not want the cause of death recorded. There are also cultural traditions around death and burial that can affect the kinds of remains that archaeologists might stumble upon many years later. For example, some societies buried babies in the walls or floors of houses. Others separated male and female graves, or buried people with certain diseases like plague or leprosy in a specific place.

Ultimately, this is a number problem. Finding three skeletons with signs of cancer in a particular area might represent 3 percent of people in a village of a hundred, 0.3 percent in a town of a thousand, or 10 percent of a group of thirty. Maybe cancer is genuinely rare in historical and prehistorical populations. Or it may have been much more common than we think, because scientists haven't systematically been looking for it. It's exciting to think about the new clues that might be revealed by DNA or protein analysis, as well as a more methodical approach toward X-ray or CT-scanning remains for signs of cancer. But what's becoming clear is that the more people look for evidence of cancer in ancient remains, the more they find.

Even though some of the most striking examples of ancient cancer have come from mummies, who have more flesh on their bones than typical skeletal remains, there's still a lot we don't know about how well tumors are preserved during the mummification process. You can't just grab a scalpel and do a post-mortem on a mummy, so researchers rely on CT scans to see what might be going on inside. But according to Casey Kirkpatrick, we don't actually know how well mummified tumors show up on a scan, so we don't know what we might be missing. To find out, she and her colleague Jennifer Willoughby decided to do an unusual experiment.

First, they teamed up with a group of researchers at a nearby hospital who had a steady supply of mice with various types of cancer. Next, they set about mummifying these animals in every way they could imagine. Some were dropped in a local swamp to replicate the mummies found in peat bogs. Others were encased in ice or ended up buried in hot sand. And for a final flourish, Kirkpatrick and Willoughby even gave a few mice a full ancient Egyptian ritual burial, carefully removing their tiny internal organs and packing the corpses with natron and natural resins before bandaging them up.* Once mummification was complete, the final stage was to put the mice into a CT scanner to see how well their tumors had been preserved through the process. Reassuringly, the signs of cancer showed up clearly in all the mummified mice, suggesting that CT scanning probably isn't missing much in the way of solid tumors when it comes to studying ancient human mummies. "Cancer is not a modern disease," Kirkpatrick emphasizes. "It has happened throughout our entire history. There are carcinogens in the environment, there are also genetic factors and infections—it's near impossible to avoid. I think we really need to reach out to the public and

* Kirkpatrick tells me they drew the line at building a miniature pyramid.

inform them about this, especially when people are suffering and thinking that cancer is all their fault."

## ALL CREATURES GREAT AND SMALL

Cancer isn't a uniquely human affliction—something I've been all too aware of since our first dog, a much-loved Welsh springer spaniel named Sheba, died of leukemia. But while it's sometimes argued that the artificial pressures of domestication cause tumors to turn up in pets as well as people (therefore putting it in the category of a "modern disease"), framing cancer as the inevitable consequence of multicellularity tells us that we should expect to see cancer occurring in any and every species. And, with some notable exceptions, that's exactly the case.

In 2014, Croatian geneticist Tomislav Domazet-Lošo and his colleagues at the University of Kiel in Germany published a mind-boggling paper describing tumors in two different species of a tiny freshwater creature called *Hydra*—the simplest organism currently known to develop cancer. Little more than a tube with tentacles, each *Hydra* is made of two layers of cells, maintained by three distinct groups of stem cells. Two of them make the layers of the tube, while the third—known as interstitial stem cells—is multitalented, capable of producing various bits of the *Hydra*'s simple body as well as germ cells, which eventually become eggs and sperm. And it's from these stem cells, interrupted somewhere along their journey to making eggs, that a tumor grows. While it's hard to tell whether a *Hydra* is feeling unwell, the presence of this cancer certainly has an effect, severely reducing their growth rate and fertility. It's also important to point out that Domazet-Lošo and his team didn't interfere with these creatures in any way, such as making genetic tweaks or putting nasty chemicals in the water—the tumors popped up entirely spontaneously. Their discovery raises an interesting question: If something as basic as *Hydra* can develop cancer, what about other animals?

One person who's trying to answer that question is Amy Boddy, Assistant Professor in the Department of Anthropology at the University of California, Santa Barbara. She and her team have been pulling together an impressive amount of data about tumor incidence across a huge range of species—a concept known as comparative oncology.

"One of the hardest things is figuring out exactly how we define cancer in the first place, especially when looking across wildly different organisms. We can be fairly sure that a cancer in a dog or mouse will be recognizably similar to a human tumor. But what about strange cells in a mussel, or a peculiar bulge on a mushroom? When you start talking about concepts of cancer in other organisms, you realize we don't know a lot about the disease," Boddy says. "There were huge arguments when we wrote our first review of cancer across the tree of life about what we should class as cancer, because the medical definition is very human-centric."

Invasive cancer in humans is defined by whether or not tumor cells have broken through the basement membrane—a thin protective sheet of molecular "cling film" wrapped around our tissues and organs. Plenty of organisms don't have this barrier layer yet can still be affected by rogue cells multiplying out of control. Plants develop large growths known as galls, usually the result of bacterial, viral, or fungal infection, or the work of wasps. And there are other strange phenomena like the fasciated cacti that we'll encounter in the next chapter.

Tumor-like masses can be found in red algae and even fungi aren't in the clear: non-invasive growths have been spotted in mushrooms, while simple molds can start growing in abnormally excessive ways. Although these lumps are a symptom of over-enthusiastic cell proliferation, it's not quite right to call them cancers as the rigid cell walls and sturdy internal structures of fungi and plants prevent rogue cells from spreading throughout the organism.

Moving on to animals, cancer turns up almost everywhere you look. One recently published list of animals known to be affected by cancer stretches over more than twenty pages, while the line-up of marine creatures that have been found with tumors reads like the menu from the world's weirdest sushi restaurant: cockles, clams, crabs, catfish, cavefish, cod, corals, and quahogs. Damselfish, angelfish, jewelfish, and goldfish. Smelt, salmon, sea bream, and weedy seadragons . . . the list goes on and on.

Tumors turn up in frogs, toads, and other amphibians, and have been spotted in a range of reptiles such as snakes, turtles, tortoises, and lizards. Cancers appear in many species of bird, from parakeets to penguins, cockatoos to cassowaries and black-bellied whistling ducks to common or garden parakeets. Not to mention the curious case of a three-legged robin with a cancerous mass in its belly, which arrived in the possession of a Mr. H.K. Coale of Chicago one day in 1919. From aardvarks to zebras, our fellow mammals are also affected by all manner of cancers: whales, wallabies, baboons, badgers, bongos, and almost everything in between.

Just as tumors turn up in long-dead human remains, there's evidence of cancer stretching way back through the fossil record. In 2003, a team led by Bruce Rothschild from Northeastern Ohio Universities College of Medicine trawled the museums of North America with a portable X-ray machine, taking images of more than 10,000 dinosaur bones. Although they found tumors in only one family of dinos—herbivorous duck-billed hadrosaurs from around 70 million years ago—they detected a staggering twenty-nine tumors in ninety-seven individuals. There's even a tumor in the leg bone of a fossilized proto turtle that roamed the Triassic seas washing over what's now modern Germany about 240 million years ago. Evidence of cancer has come to light in other dinosaur species, including a giant titanosaur, although some of these observations are controversial.*

These surveys of cancer across life have also challenged the persistently popular but incorrect belief that sharks don't get cancer. This strange idea sprang up in the 1970s when Judah Folkman and Henry Brem at The Johns Hopkins University School of Medicine in Baltimore, Maryland noticed that cartilage—the protective layer on the ends of

* As with the challenge of correctly diagnosing ancient humans, tumors in soft tissue don't get preserved. Unfortunately, fossils don't come with an accompanying veterinary report, so there's plenty of room for argument.

bones—prevented new blood vessels from growing into tumors. Shark skeletons are made entirely of cartilage rather than bones, so people started to wonder if they might be more resistant to cancer than other animals.

Lab experiments suggested that shark cartilage was very effective at stopping tumor blood vessel growth, while attempts to induce tumors chemically in these fish failed. Given that nobody had spotted cancer in any sharks in the wild, the theory seemed to check out. From there, it was a relatively simple mental leap to suggest that shark cartilage might prevent or even cure cancer. Boosted by the publication of the 1992 bestseller *Sharks Don't Get Cancer* by William Lane, a multimillion-dollar industry was born. Sharks were caught, farmed, and slaughtered in their millions to make cartilage pills for desperate cancer patients, despite at least three clinical trials showing that they were ineffective.

More importantly, the fundamental premise isn't true: tumors have been spotted in multiple species of sharks, including in the mighty jaws of a great white found off the coast of Australia in 2013. As marine biologist David Shiffman pointed out in an article about the discovery of the great white tumor, "Sharks get cancer. And even if they didn't get cancer, eating shark products won't cure cancer any more than me eating Michael Jordan would make me better at basketball."

Although shark cartilage may not be able to prevent or cure any disease, comparing cancer across species can provide useful insights into what might be going on in our own bodies. This becomes particularly interesting when we ask not *whether* a singular example of a tumor has been found in any given type of animal—something that should be entirely expected, if cancer is a process that inevitably arises in any multicellular organism—but *how often* it turns up.

Perhaps surprisingly, not only can we definitively say that cancer isn't a human-specific disease, but we also aren't even the species that gets it the most. It's a commonly held assumption that humans get cancer more than other species, yet it's based on woefully incomplete information. Just as we have no idea about the frequency of cancer in ancient human populations without systematic data collection, nobody has really looked at the incidence of cancer across species in a methodical way.

It's one thing to generate a huge list of all the species where any kind of cancer has been spotted, but it's quite another to work out whether any of them are particularly rare or commonplace. Amy Boddy and her colleagues in Santa Barbara have become animal epidemiologists, sifting through data from zoos—plus as much information as is possible to glean about wild populations—to see how common the disease really is across different species.

"Zoo animals do have unusually long lifespans compared with animals in the wild, and we have quite small sample sizes for some of these," she warns. "But our preliminary data suggests that there are quite high rates of cancer in small mammals compared with humans—we see quite a lot of tumors in ferrets, and also it looks like little mouse lemurs seem to be getting a lot of cancers too."

Boddy explains that cancer seems to be more common in animals that have been through a bottleneck—an event that drastically reduced the population at some point—meaning that today's individuals are more genetically similar than they would be if they hadn't been through the crash. Golden Syrian hamsters have squeezed through a particularly drastic bottleneck, with the majority of domesticated hamsters in the world being descended from a single litter found in the Syrian Desert in 1930. As a result, they have unusually high rates of spontaneously occurring tumors.

Other pure-bred and domesticated species are also more susceptible to cancer. Dogs have a roughly similar risk of cancer to humans, with different types of tumor being more or less common in particular breeds. And up to a third of farmed hens will develop ovarian cancer owing to the pressure of constantly having to churn out eggs.

Interestingly, humans have been through several such precarious situations in our history. For example, there's good evidence that our ancestral population collapsed to fewer than 20,000 breeding individuals around a million years ago, taking our species to the brink of extinction—something that may play a part in our susceptibility to cancer today.

Researchers have also found that "scalys"—birds and reptiles, both descended from dinosaurs—tend to get far fewer cancers compared with the furrier side of the evolutionary tree. The reason is currently a mystery, but Boddy has a few ideas.

"I think it's to do with pregnancy and having a placenta," she says, explaining that while birds and reptiles lay eggs, mammals have to maintain the capacity to generate an invasive tissue packed with blood vessels that squirms its way into the wall of the uterus, sucking oxygen and nutrients out of the mother to feed a growing fetus. Cells from the placenta and fetus also end up in the mother's bloodstream and can even become part of her normal body tissues—a process known as microchimerism. This is pretty much the exact same repertoire of biological tricks that a cancer needs to grow and spread, and many tumors even hijack the same genes and molecules to gain a foothold in the body.

There was an idea kicking around for a while that mammals with more invasive placentas—including humans—might be more susceptible to cancer than those with more superficial systems, such as horse or cow, while cats and dogs have placentas and cancer risks that fall somewhere in between. Frustratingly, this neat theory doesn't seem to check out now that Boddy and her team have gathered more data from different species. And she's still lacking information on cancer rates in placenta-less marsupials, which give birth to tiny live young and grow them in a front-loading pouch, or egg-laying monotremes like the duck-billed platypus. Even so, she's convinced that there's a connection between a species having the ability to grow a placenta and having a greater likelihood of developing cancer.

"I think there is a connection between the two," she says, pointing out that a fetus is made of cells that are genetically similar but not identical to its mother—a situation that risks triggering a lethal rejection by the immune system. "We could have evolved to keep everything contained in the uterus, but we have evolved this placenta that invades and integrates into every single tissue in the maternal body, so I think mammals are potentially not as good at being sensitive to detecting tumors that are a slightly mutated version of ourselves."

## SIZE MATTERS

There's an even more intriguing curiosity about cancer incidence in humans and other species, and it goes like this: If cancer is an inevitable consequence of multicellular life, and is likely to arise in any population of cells, then it should follow that the more cells in an animal, the more likely it is to get cancer. More cells equals more cell proliferation equals more chance that something will go wrong. With great size comes great risk, and this problem should be exacerbated in animals that live for a very long time.

"We know that within a species, the bigger the individual the higher the cancer rate—for example, taller, bigger humans have a higher risk than shorter, smaller ones, and the same is true for dogs," Boddy explains. "You can think about it just as being a question of probability due to having more cells, but there's also potentially sexual selection. If you grow big quickly, you're entering the mating game sooner."

By way of example, Boddy tells me about the mating habits of Platy fish—colorful little creatures that are native to Central America and found in aquariums all over the world. Some male fish carry a gene fault that makes them grow unusually large, which is particularly attractive to females. Unfortunately, the same mutation also makes the males susceptible to developing melanoma. By the time they develop a cancer that hampers their health it's too late: they've already matured and mated, passing their rogue gene on to the next generation.

There's a similar story with white-tailed deer. Males invest a lot of time and testosterone in growing impressive antlers (the bigger the better as far as females are concerned). This effort comes at the cost of increased risk of developing antleromas—fibrous tumors that press into the skull and damage the brain, even leading to death.

Here's where it gets strange, though. Although the relationship between larger body size and increased cancer risk holds up if you compare individuals of the same species, it vanishes once you look more broadly across the tree of life. Large, long-lived animals like whales and elephants actually have fairly similar rates of cancer to small, short-lived creatures like mice. This is a remarkable observation given that a 200-ton blue whale is 10 million times bigger than a ¾-ounce mouse, meaning that a mouse-sized piece of blue whale flesh must be at least 10 million times more cancer-proof than an actual mouse.

Humans are a clear outlier, with higher than expected rates of cancer for our size. But when you take our bad habits out of the equation (particularly smoking), then we seem to be remarkably resistant to cancer compared with smaller creatures but much more susceptible than the giants of the mammalian world. The observation that cancer risk doesn't track with body size is known as Peto's Paradox, named after Richard Peto, the British statistician who first noticed it back in 1976. While it may seem contradictory, this eponymous Paradox is a fascinating lens through which to think about why humans—or any other organism—might or might not get cancer at a given point in their lifetime. And all that's required to solve it is a little strategic thinking.

As well as differing in size, animals also differ in lifespan. In the wild, under constant risk of predation, a mouse might be lucky to make it to a year. Even in the cozy confines of a laboratory, the oldest are lucky to make it past two. By contrast, the Greenland shark—the oldest known vertebrate—reaches sexual maturity at the ripe old age of 150. Using a dating technique that looks for the impact of 1950s radioactive bomb tests in the lens of the eye, the oldest of the Greenland sharks tested so far was thought to have been alive for up to 500 years, first slipping through the chilly Arctic seas when Queen Elizabeth I was on the throne. African elephants average about sixty to seventy years, but guinea pigs are unlikely to reach eight. Global human average lifespan is now around seventy, while our chimpanzee relatives can expect to hit fifty. On the other end of the primate spectrum, mouse lemurs have an average reproductive lifespan of around five, although they can live to fifteen in a zoo.

Solving Peto's Paradox requires an evolutionary trade-off between growth, longevity, and sex. To put it simply, either you evolve to live fast and die young—existing for just a few short, dangerous years filled with as much reproduction as possible—or you're a slow burner that grows large, tends to eat rather than be eaten, has offspring later in life and looks after them for a long time.

Obviously, if humans all got cancer before any of us could reproduce, we wouldn't have got very far as a species—that's how natural selection works. But maintaining a large body in a healthy, cancer-free state for decades takes a lot of energy and resources, so species have evolved to stay healthy for the duration of their reproductive phase, however long that may be, succumbing to cancer when the effort to maintain the body is no longer worth it. It therefore makes perfect sense that 90 percent of all human cancers occur in people over the age of fifty: we've evolved to get through the prime of life in good health, but once the kids are born and raised then all bets are off.*

The ultimate expression of the "live fast, die young" strategy comes from a marsupial mouse known as *Antechinus*. For around two weeks in August, in the depths of the Australian winter, males will mate with as many females as possible in frantic bouts lasting up to fourteen hours. But as the mating period draws to a close, bad things start to happen to the little guys. Their fur falls out, their internal organs start to break down and infections quickly take hold. Within a few short weeks all the males are dead, having invested all of their energy in reproduction and literally gone out with a bang.

Their mates fare little better and the mothers usually die after their pups are weaned, leaving the orphans to fend for

* There's an intriguing difference in cancer incidence between the sexes that seems to be independent of anything else, with men being slightly more likely to get cancer at a younger age. It's a controversial idea, but the so-called Grandmother Hypothesis suggests that while grannies are useful because they help out with the grandkids, grandpas who take less of a role in childrearing are more dispensable in evolutionary terms.

themselves until the following year, when the whole cycle starts again. The reproductive strategy of these creatures may seem strange in comparison to our human lifestyles, but to them it makes perfect evolutionary sense. *Antechinus* live on insects, which tend to appear in cyclical gluts. The frenzy of mating happens just around the time of the biggest food bonanza, so the mothers are well fed as they suckle their young, while the males are little more than disposable sperm-delivery vehicles.

On the other end of the spectrum, researchers studying nature's slow burners are making some intriguing findings about how these species manage to stave off cancer for so long. Advances in DNA sequencing mean that we can now rummage around in these animals' genomes and find out what keeps them ticking.

One of the most famous examples of a long-lived, cancer-resistant mammal is the naked mole rat. These sand puppies live in large colonies beneath the African desert, constantly tunneling in an effort to find tasty plant roots and keep their continually growing teeth in check. Shielded from the sub-Saharan sun, their burrows stay at a constant 86°F, so they've dispensed with the effort of maintaining the high body temperature common to all other mammals. They don't seem to feel pain, can survive in perilously low oxygen levels, aren't pestered by predators, and rarely venture out into the scorching sunlight. Even more strangely for rodents, they're eusocial; only a few animals in the colony are sexually active—a sole dominant queen who rules the roost and a handful of lucky stud males—while the rest are non-breeding workers responsible for digging, maintaining, and guarding the twisting network of tunnels.

Although researchers initially became interested in naked mole rats for their unusual social structure, they soon realized something odd about the animals they'd brought into their captive lab colonies: they just weren't dying. In 2002, researchers in New York published a report of a naked mole rat in their laboratory colony that had made it to at least twenty-eight, beating the previous rodent longevity record holder (a 27-year-old porcupine). This was smashed in 2010 by a naked mole rat nicknamed Old Man, who was thirty-two when he finally passed on to the great colony in the sky. Most mole rats make it into their late twenties and cancers are virtually unheard of, with only a handful of cases documented in more than a thousand captive animals.

It's still not entirely clear how naked mole rats manage to live so long and stay cancer-free. Maybe it's their low-calorie, low-temperature lifestyle, which is thought to reduce the production of damaging chemicals called free radicals that are created as cells generate energy. Perhaps the explanation lies in altered levels of hormones and other molecules that drive cell growth, or in their polyphenol-rich vegetarian diet. In 2013, scientists discovered that the mole rats make an unusually large and sticky version of a kind of cellular glue called hyaluronan. They suspected that this helps to reinforce contacts and communication between their cells, stopping them from running out of control and turning cancerous.

Certain genes involved in energy production are much

more active and present in many more copies in mole rats than in mice. Perhaps this extra dose of DNA acts to buffer the carcinogenic effects of genetic damage, keeping mole rats powering through well into old age. There are other key differences in genes involved in the response to DNA damage and other aging-related pathways. Cells from naked mole rats are more resistant to stress and damage than those from other small rodents. And a study published in 2019 showed that naked mole rats also have a highly unusual repertoire of immune cells compared with mice, which may help to keep them healthy for so long.

As if that wasn't enough, they have yet another layer of protection against excessive cell growth: it's simply not tolerated. There's a phenomenon in biology known as contact inhibition, best described as a cell's "personal space," which stops them proliferating if things start to get too crowded. Naked mole rat cells are exquisitely sensitive to contact inhibition and will freeze as soon as they detect another cell getting too close, preventing any pile-ups that could herald the start of a tumor.

Similarly long-lived blind mole rats (no relation to the naked ones) have solved Peto's Paradox in a different way. Although these rodents are about the same size as regular rats, they live five times as long and have very low cancer rates, often making it to their twentieth birthday. This longevity seems to result from the ability of blind mole rat cells to repair potentially cancer-causing DNA damage five times more efficiently than those of a normal rat—a property that might have evolved to protect the animals from the harmful cycles of high and low oxygen levels that they experience in their subterranean burrows.

Capybaras—those delightfully chilled South American giant guinea pigs with a reputation as being the friendliest creatures in the zoo—have a different answer. Their unusually large body size appears to be the result of overenthusiastic activity of the hormone insulin, which controls cell growth and metabolism. At the same time as becoming King of the Rodents, they must have also evolved a way to suppress cancer (remember, the bigger the body, the more cells and the greater the cancer risk). Researchers digging in the capybara genome have recently found that although these animals seem to have higher than normal levels of harmful genetic mutations compared with other rodents, they also seem to have particularly vigilant immune cells that seek out and destroy rogue cells before they can grow into a tumor.

Elephants are a whole other story. Rather than attempting to fix any potentially cancerous damage to their DNA or ramp up their immune system, they've evolved multiple copies of a gene encoding a molecule called p53—the so-called Guardian of the Genome—which triggers cells to die for the greater good at the first hint of trouble. This makes sense considering their huge size: you've got cells to burn if you're an elephant, so it's better to get rid of any suspicious ones straight away.

Scientists have also been diving deep into the genes of the 100-ton bowhead whale, whose 200 years of relatively cancer-free living make it a good candidate for the longest-lived mammal on the planet. At the moment it's not clear how they achieve this, but it may be to do with having gained or lost certain genes associated with DNA damage repair or by controlling cell proliferation.

On the other end of the size scale, a teeny Brandt's bat clocks in at less than half an ounce—one ten millionth the weight of a mighty bowhead and about half the size of a typical laboratory mouse. Yet it holds the record for lifespan in a creature that small, with the oldest documented individual living for a staggering forty-one years. While Brandt's bats are the winners of the old age Olympics, all other species of bats also live for an unusually long time compared with ground-based rodents of similar sizes. There's certainly a built-in longevity advantage to being able to fly, as bats can zip off at the first sign of predators. But they seem to have some useful molecular adaptations, too.

In 1961, microbiologist Leonard Hayflick realized that most cells could only divide fifty or so times before they ran out of steam and died. We now know that this Hayflick limit is imposed by telomeres—the caps of DNA and protein on the ends of chromosomes that protect the fragile ends in the same way the plastic aglet wrapped around the end of a shoelace stops it from fraying. In most normal animal cells, the telomeres get a tiny bit shorter every time a cell divides, owing to the vagaries of the DNA copying machinery. Once the telomeres shrink to a certain size, then the cell will die. However, embryonic stem cells crash straight through the Hayflick limit, dividing prolifically as they create all the tissues of the body during development. To avoid a chromosomal crisis, they activate a gene encoding an enzyme known as telomerase, which rebuilds telomeres to the correct length during every cell division.

This molecular "countdown clock" acts as a natural cancer defense mechanism, preventing cells from multiplying out of control. In fact, reactivating telomerase and resetting the telomere clock to enable endless, immortal proliferation is a vital step in the journey toward cancer. Intriguingly, the telomeres in the longest-lived species of bats don't get shorter with age, so they can keep on repairing their tiny bodies for decades. Sidestepping the relentless tick of the telomere clock doesn't seem to increase their risk of cancer though, suggesting that there must be other as yet unknown antitumor mechanisms at work.

One of the more left-field theories about the reduced risk of cancer in large animals proposes the idea of hypertumors. These are "super-cancers" that arise within the already lawless environment of a tumor and start destroying the bad cells that are already there. This concept of a tumor within a tumor may sound strange but, as we'll see later on, given that every cancer appears to be a patchwork of genetically unique clusters of cells, it's certainly possible that cellular infighting might help to suppress tumor growth up to a point.

There also seems to be a connection between cancer risk and the body's healing ability. Animal cancer expert Amy Boddy told me about the time she went to her collaborator Tara Harrison at San Diego Zoo in search of some skin cells. Most zoos are happy enough to provide small skin samples from most of their charges, collected under local anesthetic using a small hole punch-like device. But when it came to one particular animal—the giant Galápagos tortoise—the answer was a flat no. Deliberately punching a hole in the skin of one of these gentle giants creates a wound that takes more than a year to heal, compared with a week or less for most other animals in the zoo—something that the tortoise keepers can't justify on animal welfare grounds.

It's certainly thought-provoking to compare the slow healing of a cancer-resistant tortoise, with its wrinkly, thick hide and protective shell, with the swift sealing of soft human skin that can be slashed with something as pathetic as a piece of paper. Mice heal even faster still. But evolving the ability to heal rapidly also means that cells quickly have to be able to flip into proliferation mode, increasing the chances that one of them might go rogue. In terms of evolutionary strategies, humans and mice have opted for supple skin and speedy healing, but at the expense of a layer of potential cancer protection.

Different species have solved Peto's Paradox in different ways, each employing their own particular strategy to get through their reproductive years in one piece. And there are still plenty of things we can learn from studying organisms that set off on a different evolutionary path from us millions of years ago.

## CANCER-PROOF

Despite the seeming ubiquity of cancer across every branch of the tree of life, there are a few animals that just don't get cancer at all, as far as anyone can tell. Comb jellies are one such lucky species—transparent torpedo-shaped creatures that shimmer with iridescent rainbows diffracting through the rippling fronds they use to propel themselves through the sea. Although comb jellies can range in size from just a few millimeters up to five feet, there are currently no known examples of cancer in any of the 100 or more species that have been described to date.

Another example is *Placozoa*. These elusive aquatic organisms are considered to be the simplest multicellular animals in existence and are little more than a blobby collective of a few thousand cells, made up of just four distinct types. It's tricky to tell exactly what a *Placozoa* tumor might look like, but they don't seem to get cancer. They also have an unusual way of resisting carcinogenic X-ray damage by simply popping clumps of damaged cells out from their surface in the same way that you or I might squeeze a spot.

Finally, there are the sponges. Carlo Maley, Director of the Cancer and Evolution Center at Arizona State University in Tempe, takes me into his lab to look at a salty tankful of prickly white spheres, each about the size of a Mentos. These are *Tethya wilhelma*—just one of the many species of sponge that appear to be untroubled by any form of cancer.

"We wanted to find a new model organism that would be good to study, where its genome has already been sequenced, and we can grow it in the lab," Maley tells me, explaining how one of his colleagues, Angelo Fortunato, has spent months of his life setting up the perfect saltwater system to keep these sponges happy in their new home. And then, after all that, he's bombarding them with X-rays.

This is no gentle zapping—it's a full-on nuclear assault.

By way of comparison, a short, sharp dose of just five grays of high-energy radiation is enough to kill a human within two weeks of exposure. Fortunato's giving his sponges a staggering *seven hundred*, and they just carry on like nothing happened. No obvious signs of harm and certainly no cancers.

Maley and his team are busy figuring out exactly how these super-sponges manage to shrug off such a colossal blast, in the hope that it might reveal new insights into how to protect our own cells from radiation damage. This might be useful when it comes to finding ways to enhance the fatal effects of radiotherapy on cancer cells or to protect healthy tissue around them. At the time of writing, they're still searching for clues, although other researchers have found a number of chemicals in sponges that block tumor growth. These little marine creatures definitely have something interesting going on inside their unassuming exteriors that needs to be wrung out.

## MODERN LIFE IS RUBBISH

Cancer is neither new nor a uniquely human disease, so we can't blame it entirely on the evils of modern living. But we should question why the rates are so high in affluent societies, with one in two people in the US born after 1960 predicted to get cancer at some point in their lifetime. Some of this can be explained by the impressive increase in life expectancy—more and more of us are simply living long enough for cancer to kill us in old age rather than succumbing to violence, predators, misadventure, infectious diseases, starvation, or death in childbirth.

Nineteenth-century doctors were convinced that cancer was a disease of "civilization" but, as we've seen, it's difficult to generate accurate figures for cancer in ancient populations. Gathering stats on more recent hunter-gatherer societies and current populations with less "modern" lifestyles is also a major challenge. Countries like the UK have incredibly detailed cancer statistics, fed by the detailed medical records kept by the National Health Service. It's therefore unlikely that anybody would die from cancer in the UK without it being noted at some point. But there are still many parts of the world where cancer goes undiagnosed and unrecorded.

Humans are immensely adaptable to the changing world around us and our genes are changing, too. We can see evidence of genetic changes that have swept through populations at a relatively quick pace, such as the ability to digest milk after infancy owing to a specific genetic change that became widespread during the rise of dairy farming around 10,000 years ago. Genetic variations for blue eyes are also fairly new, popping up between 6,000 and 10,000 years ago. But the world we live in today is changing far faster than this.

Our modern bodies evolved in a world with an uncertain food supply and more physical activity, and probably a different suite of infectious diseases and exposures to carcinogens. Ancient humans may have been exposed to indoor fires and chemicals from tanning or smelting, but they didn't deliberately inhale cigarette smoke or roast themselves in the midday sun. Our lifestyles are very different, too. For example, women in more developed countries tend to have fewer children and breastfeed for a relatively short time, as well as starting their periods earlier and having the option of hormone replacement therapy during menopause. Given the role of hormones in driving the growth of many breast cancers, it's not a huge logical leap to suggest that altering the balance of hormones over a lifetime might have an impact on risk.

All this talk of evolutionary strategies makes me wonder whether or not humans will gradually start to evolve our way out of cancer as our lifespans increase and the average age at which we have children also rises. Disappointingly, every scientist I've asked tells me that this is wishful thinking. Evolution works on a timescale of millennia, not centuries, and there simply hasn't been enough time for our species to adapt to all the changes that have happened. There's nothing we can do about the slow tick of time working in our tissues, honed by hundreds of thousands of years of natural selection.

As I head into my forties, I'm increasingly aware that I'm approaching the age where evolution gives up on me. I can do my best by not smoking, watching my weight, being careful what I eat, taking care in the sun, and cutting down on alcohol, but ultimately I'm trying to wage war against my biological destiny.* Even so, as I talk to Amy Boddy about her research menagerie, I'm excited to think that this broader view of cancer could reveal important truths about the

* You can't cheat this destiny, either. Although evolution has got humans to a point where we're unlikely to get cancer during our prime reproductive years, you can't dodge it by not having children.

disease; although it feels like the whole field of comparative oncology is only just getting started.

"I think we need a better understanding of cancer in other organisms and how important that could be in revealing the basic biology of the vulnerabilities in humans, and I'm also sad to see that there isn't more work looking at human variation across the world in different populations and small-scale societies compared with Western populations," she says. "This is nature's toolbox, giving us all the recipes and ingredients that it has developed over millions of years of evolutionary history to produce different cancer defense mechanisms and modify risk—evolution has given us a pretty good code for what works."

There's one final reason why we should care about cancer in animals as well as humans and that's for the sake of the animals themselves—an argument that often gets lost in the anthropocentric world of cancer research. Boddy is passionate that we should care about why animals get cancer as much as we should care about why humans do. For a start, vets and conservationists are keen to discover more about cancer in domesticated, captive, and wild animal populations, both to understand the causes and to figure out how best to treat it. And the unexpected appearance of cancer in animals living in a particular area might reveal the presence of carcinogens that their human neighbors would do well to avoid, too. We could even learn something from our animal friends along the way, perhaps stealing their best anti-cancer innovations to apply to our own failing bodies.

Still, knowing that cancer has always been with us and

affects almost every branch of the tree of life still doesn't tell us *why* it happens. What turns a normal, well-behaved cell into a nasty one that grows out of control and causes trouble? To understand that, we need to uncover the rules governing the cellular societies within all living things and see what happens when they get broken.

# 2

# THE PRICE OF LIFE

Back in the early days of life on Earth, every cell was its own entity—an island in a sea of other free-living cells. But after a billion or so years of the single life, it was time to settle down. Cells began to club together and communicate with each other, forming small multicellular organisms. At first these were little more than loose collectives, but over millennia they evolved into highly organized creatures. They learned to specialize and differentiate their many parts, forming distinct tissues and organs: a place for every cell and every cell in its place.

Cells have decided that it's better to buddy up and form multicellular organisms than go it alone at several points during the history of life, creating the progenitors of fungi, algae and plants. Multicellular animals are thought to have evolved only once, first appearing on the scene around 600 million years ago. Although becoming multicellular means that each individual cell loses its autonomy, only replicating exactly when and where is necessary—during development, growth, or repair, for example—there are some big advantages to being part of a larger whole.

For a start, multicellular organisms can grow large, providing a significant survival advantage (because it's hard to get eaten when you're bigger than everything else around you). They can eat a wider range of foods and evolve adaptations to cope with a range of environments, moving further and faster than unicellular slowpokes. Having lots of cells also means that specific tasks can be allocated to particular parts of the body—known as differentiation—allowing much more sophisticated functions to emerge than would be available to a single-celled jack of all trades, such as nerves, muscles and blood. Cells within a larger organism can also work together to create "public goods"—things like nutrients or other chemicals that are needed for growth. If you're a single cell living on your own, anything you produce will spill out into the environment around you where it can be gobbled up by your competitors. But products that are made within a multicellular body stay inside, benefiting the whole organism and helping it to grow.

Most excitingly, if you're multicellular it means that you can have sex, rather than just handily splitting in two like bacteria when it's time to reproduce. The evolution of sex in multicellular animals has led to a clear distinction between the cells that make eggs or sperm, formally called germ cells, and those that make up the rest of the body, known as the soma or somatic cells. The core purpose of the soma is to get on with the dirty work of staying alive—feeding, fighting, finding mates, and so on—while the germ cells are carefully protected in order to pass on the genetic torch to the next generation.

A multicellular lifestyle only works if there are tight controls on cell division and function. Single-celled organisms like bacteria have one evolutionary goal: to proliferate and pass on their genes. A deceased unicellular organism is literally an evolutionary dead end, so there's a big incentive to stay alive and keep on replicating. But in a multicellular organism, proliferation is allowed only as long as it's part of normal development and growth from baby to adult, to heal wounds, or as part of the regular running repairs required to maintain the body. Cells also need to make sure they stick to their designated roles: a neuron in the brain can't suddenly decide to produce insulin like the islet cells in the pancreas. Cells in your skin need to stay put, forming an impermeable barrier against the world rather than crawling off on a journey to another part of the body. And any malfunctioning or damaged cells should die or be picked off by the immune system instead of hanging around to cause trouble.

Multicellularity is therefore best seen as a biological social contract, with each cell doing its duty for the greater good of the organism. Cancer cells ignore these rules, proliferating out of control and invading the surrounding tissue, eventually spreading through the body and ultimately resulting in death if they can't be successfully controlled. In order to understand where cancer comes from, we first need to understand the rules of multicellular life and what happens when they're ignored.

## MEET THE CHEATING AMOEBAS

The soil-dwelling slime mold *Dictyostelium discoideum* (*Dicty* for short) spends its days mooching about in the soil as a single-celled amoeba, as long as life is good and there are plenty of tasty bacteria to eat. But when food supplies run low, the solitary cells start sending out SOS signals that cause them to cluster together. Up to 100,000 cells will gang up to form a small, slimy blob just a few millimeters long—unimaginatively known as a slug—which slithers off in search of a nice brightly lit spot with the right temperature and humidity. Once there, the slug transforms again. This time it pushes up a vertical stalk topped with a bud-shaped fruiting body. Finally, this bud bursts open and scatters tiny spores far out into the world in the hope of finding more amenable conditions, each of which can germinate into a new *Dicty* amoeba and start the cycle again.

The life cycle of *Dicty* appears to be a shining example of the benefits of multicellularity, with individual cells teaming up to reproduce when the going gets tough. It also highlights the downside of being part of a cellular society: although 80 percent of the cells that make the slug will end up becoming spores and get another shot at life, the remaining 20 percent that end up in the stalk will die, sacrificing themselves for the greater good of the colony. But even in this simple society there are cheats that break the rules.

In 1982, Yale University biologist Leo Buss noticed some antisocial behavior going on in the world of slime molds. He saw that particular cells in a related species, *Dictyostelium* *mucoroides*, were more likely to end up in the fruiting body than in the stalk, giving them a much better chance of surviving and passing on their genes. He called these cheats "somatic cell parasites."* A quarter of a century later, Gad Shaulsky and his colleagues at Baylor College of Medicine in Texas published a paper showing that the same selfish behavior also happened in *Dicty*, resulting from alterations in any one of more than 100 different genes.

Then they saw something even more curious: cheaters only cheated when they were surrounded by amoebas they weren't directly related to, with some families of cells only making a paltry 5 percent contribution to the stalk in the presence of genetically different neighbors. But when they were surrounded by genetically identical progeny, the full 20 percent of cheats knuckled down and accepted death—laying down their life for their family but not for random strangers. Presumably there's no extra benefit in muscling to the top of the stalk if the family genes are going to be passed on by their kindred anyway.

We should be careful not to imply agency or intelligence on behalf of these slimy cheats. They're merely responding to their genetic programming, which has been shaped by natural selection. Once a genetic variation arises in an amoeba that makes it more likely to push to the top of the stalk, that cell is more likely to survive and keep on multiplying, creating a new generation of cheats that also carry the same mutation. But it's amazing that an organism as simple as a single-celled

* I like to think of them as "asshole amoebas!"

slime mold contains so many genes that maintain multicellular social behavior. And it's even more incredible that these social "rules" go out of the window once those genes are altered, but only if it's evolutionarily advantageous to do so: a whole population of selfish amoebas would come unstuck pretty fast if there weren't any individuals prepared to create the sacrificial stalk.

Scaling up from *Dicty*, the world is full of endless examples of cheats and scumbags who break the rules of their society. In the 1970s, a group of mathematically minded evolutionary biologists coined the term "sneaky fuckers" to describe the behavior of young red deer stags. Having failed to secure their own harem of females, they'll nip in to mate while the older, larger males are busy fighting over ownership of the hinds. Genetic testing has since proved that a surprisingly high number of these stealth encounters led to the birth of baby Bambis, suggesting it's a remarkably successful mating strategy. The same reproductive gambit has since been discovered in many other animal populations.

Then there's the example of the Cape honey bee. Like most social insects, these busy bees live in a strictly hierarchical colony divided into female workers and male drones, ruled over by a queen. She's the only female in the hive that gets to mate, producing powerful hormones to suppress the sexual urges of the workers. If the queen bee goes AWOL, workers can reactivate their ovaries and start laying unfertilized eggs that hatch into male drones. But, just occasionally, Cape honey bee workers rise up in rebellion, rearranging their reproductive processes to produce female offspring and regal pheromones even if a queen is already present (a phenomenon known as thelytoky from the Greek words *thelys* and *tokos*, meaning "female birth").

Being able to activate Queen Mode allows regular workers to become cheaters, ignoring their usual jobs to lounge around producing baby bees. Groups of fake queens will invade neighboring hives of a closely related subspecies of Cape bee, taking over from the hapless queen and workers inside, and cranking out even more pseudo queens. As the hive gradually fills up with the offspring of these cheating queens, there are fewer and fewer workers left to get on with the important business of collecting nectar and pollinating plants, eventually resulting in colony collapse.

Remarkably, a team of South African and German researchers recently discovered that the ability to become a cheating queen comes down to a change in a single "letter" in the bee genome, located within a gene whose function is currently unknown. These cheats are remarkably prosperous and are widespread across the north-eastern region of South Africa, despite bringing about the demise of their hive and causing misery for local beekeepers.

Even so, from an evolutionary perspective this ability to repopulate a hive with female workers and a new queen is extremely useful. It's very windy in the region of South Africa where the bees normally live and queens have a tendency to get blown away when they venture out of the hive. Under these blustery conditions, tolerating the risk of rogue queens seems like a small price to pay to ensure the overall survival of the species.

## A THORNY PROBLEM

By early May, Arizona State University's Tempe campus is already baking in dry 100°F heat that makes your eyeballs shrivel and your skin crawl. It's a bad place for a pale British writer who's prone to sunburn but a perfect home for a cactus. The latest addition to the university grounds is a small collection of crested cacti, nestled in a gravelly bed between two faculty buildings. These are no ordinary plants. Rather than poking neatly rounded fingers toward the sky, their stems explode in an exuberant array of swollen growths. It's impossible to look at the riotous, prickly lumps and not spot the similarities with cancerous tumors bulging inside a human body.

The parallels are obvious to Athena Aktipis, the woman who planted them there. She's the director of the Cooperation and Conflict lab at Arizona State University and co-director of the intriguingly named Human Generosity Project—a massive interdisciplinary research program studying societies and cultures all over the world in search of the forces that shape human generosity. After completing a PhD on the evolution of cooperation in human societies, she became intrigued by the idea that the principles of functional (or dysfunctional) societies might also work down on the cellular level. But it wasn't cancer that first got her interested in the concept of cellular society—it was a crested cactus.

"I found this website that had all these amazing pictures of crested cacti," she tells me as we sit in her office hidden deep within the ASU's Psychology Department. "There's something really profound about the fact that cancer-like phenomena are happening not just in animals, but also in life that on the surface looks so different from us. Plants are biologically alien to how we think about ourselves and other animals, but that switched on something for me about how cancer is a very fundamental thing about life."

Rather than getting caught up in cells, molecules, and genes like most other cancer researchers, Aktipis wondered whether her theories about how individuals in a society cooperate might bring a different perspective. Looking back over her work framing societies as networks of individuals with shared resources and responses to challenges, she suspected that the organized tissues of the body must normally act as cooperative societies of well-behaved cells, all adhering to five golden rules: don't over-proliferate; do the tasks you're meant to do; don't take more resources than you need; clean up after yourself; and die when you should.

Just as these rules will enable any society to function well—including our own—problems will arise if individual members decide to go their own way. Cancer cells cheat by breaking all these rules, initially maybe one at a time but then all at once as they take hold and spread through the body. They multiply out of control, ignore their normal function within an organ, gobble up oxygen and nutrients, create a toxic, acidic environment, and steadfastly refuse to die after they ought.

Multicellular organisms have evolved over a billion years to function as societies of cells, with every unit working in its specified role toward the common good and the propagation of the species rather than the needs of an individual cell. This rigid hierarchy leaves no room for the free-and-easy lifestyle of our single-celled forebears. Cell division is tightly controlled, dictated by a host of complex, intertwined molecular and genetic pathways, ensuring that a cell only divides exactly when and where it's needed. Disorder will not be tolerated. There's no space for damaged or disobedient cells. Troublemakers are encouraged to commit suicide for the good of the rest. Old cells are peacefully put to sleep. Strict as it seems, this regimen is what keeps us healthy.

Yet within any organized society—human, animal, or cellular—there will always be individuals who bend the rules (and I'm sure we've all done it ourselves, especially if we know we can get away with it). Just as human societies thrive and grow best when people cooperate and have social or legal norms that limit competing or cheating, the evolution of stable multicellular organisms also requires the suppression of cellular cheating. The more cells there are—and the longer they have to hang around—the more challenging that becomes. A huge amount of effort has gone into suppressing cheaters during the evolution of multicellular animals. The bigger you are, the more members in your cellular society and the greater the likelihood of cheats emerging, so you'll need more control mechanisms to suppress them.

For an individual cell, committing to being part of a bigger multicellular project means giving up autonomy and no longer being in charge of your own evolutionary destiny. Instead, you buy in to the hope that the body you're part of manages to pass on your shared genes before it dies. But there's the ever-present incentive to break the rules, throw off the oppressive shackles of your cellular society, and just start proliferating anyway.

Unfortunately, an obvious problem quickly arises. Cheating shifts the balance between the long-term goal of the organism—to live long enough to reproduce—and the internal drive of the cheater to profit personally over its neighbors in the short term by growing into a malignant cancer, even if that's ultimately at the expense of its host. There's also a natural limit to the number of cheats that any society can tolerate: the organized body of a multicellular animal would devolve into chaos just as human society would quickly descend into a *Mad Max*-style dystopia if everyone decided to cheat at once.

## COME TOGETHER

Iñaki Ruiz-Trillo, leader of the Multicellgenome Lab at the Institute of Evolutionary Biology in Barcelona, Spain, is a man entranced by an amoeba. More specifically, he's obsessed with *Capsaspora owczarzaki*—a single-celled organism that's thought to be the closest known relative of multicellular animals, thanks to its unusual lifestyle.

Unlike most other unicellular beings that keep themselves to themselves, *Capsaspora* goes through three stages of life. It can be found as single cells crawling around inside the blood of freshwater snails. It can also form compact spore-like cysts. But the third type is the most curious. In response to an as-yet-unknown signal, the amoebas will crawl toward each other, clumping together to form a little gang, even producing a strange glue that holds them together. It's here, in the grey area between uni- and multicellularity, that we can start to figure out the rules that govern multicellular life and begin to understand how cancers break them.

There are plenty of surprises that Ruiz-Trillo and his team have found within *Capsaspora*'s genes that might help to explain the origins of multicellular societies as well as the roots of cancer. As might be expected, *Capsaspora* has a full suite of cell cycle genes that enable it to replicate, along with all the other machinery that's required to build a cell and make it work, such as switching genes on and off or generating energy. But then there are some other genetic bells and whistles that, at least on the surface, seem totally unnecessary.

Curiously, almost all the same genes and molecules that multicellular animals use to build the different tissues of the body are present in *Capsaspora*. In fact, Ruiz-Trillo has spotted almost every single invention of multicellular life in this peculiar unicellular ancestor. For example, *Capsaspora* cells make molecules called integrins, which are found on the surface of animal cells and enable them to stick together and form organized structures. *Capsaspora* also have their own special versions of genes that were previously thought to be exclusive to animal development, enabling cells in the early embryo to make fundamental decisions about which way is up or down, front or back, left or right.

Then there are four or five genes that look like they're part of a collection of genes known as the Hippo pathway, which controls how big an animal's body grows. Ruiz-Trillo can even take these Hippo genes out of *Capsaspora* and put them in a fruit fly, where they take over control of the size of the fly's eye. But amoebas don't even have eyes (or any other organs), so what does a unicellular organism need with all the molecular trappings of a complex multicellular animal? "All of these mechanisms work as Capsa transitions from one stage to the next—from amoeba to cyst to aggregate," Ruiz-Trillo explains. "It's just the same process as differentiation within a more complex animal, but separated in a temporal way—you become one thing then you become another thing then you become another thing—but each cell can only be one thing at any one time. This is the big problem of being a unicellular organism."

A baby growing in the uterus develops hundreds of different cell types, each uniquely specialized to a particular task. But if we were like *Capsaspora*, we'd go through a stage when our entire body just made liver, then switched to brain or muscle. As Ruiz-Trillo's work shows, the main thing that differentiates humans and other more complex organisms from a loose collective of amoebas is that we've evolved such that our cells can be allocated to do different jobs at the same time. But although *Capsaspora* has all the genes it needs to create different functions, it's not terribly good at multitasking.

More complex organisms have evolved multiple layers of complex mechanisms to control when and where their genes are switched on. While many of these controls are still present in *Capsaspora*, it's missing the myriad long-range genetic "control switches" scattered throughout the genomes of more complex multicellular creatures. It's these that turn genes on at the right time and in the right place during development to create all the different tissues of the body. Intriguingly, it's these kinds of long-range interactions between switches and genes that frequently get messed up in cancer cells.

As well as lacking the genetic sophistication of multicellular organisms, there's one more crucial feature that *Capsaspora* appears to lack: death. Although these tiny amoebas can and do die, they don't appear to have the crucial components of a cellular "suicide" program, known as apoptosis, which kicks in when cells are damaged or no longer required.

In multicellular organisms, including humans, apoptosis provides powerful protection against cancer. You can even see it at work in your own body if you're careless enough to get sunburned: those flakes of dead skin that peel off a few days later are damaged cells that are too messed up to replicate properly, so they've been instructed to die rather than hang around and potentially cause problems later down the line. Unsurprisingly, genetic alterations that disrupt the key components of this suicide response are an essential step in the development of cancer—not only does a growing tumor require cell proliferation to make new cells, but it also needs those new cells to *not die*.

This discovery highlights the contradictory pull between the needs of each individual in a multicellular body and the greater good of its society. The "aim" of a single-celled organism is to make more of itself and avoid dying, but from the point of view of a multicellular being, the most important thing is that cells die when they're damaged, faulty or no longer necessary, for the greater good of the group.

"For billions of years cells have just divided—if you're

unicellular, that's just what you do. But once you're inside another organism, it's no longer all about you," Ruiz-Trillo tells me. "Being multicellular creates a lot of benefit for the organism and you benefit too, but now you're in a very different situation, so some of the things that you were able to do on your own you can't do any more because you have to behave and follow the rules."

It's a problem that's immediately familiar if you've ever been on holiday with a group of friends. Activities or decisions that seem simple enough as a singleton suddenly become fraught battlegrounds of desires and priorities, all pulling in different directions. As Ruiz-Trillo explains, a lot of the students in his lab like to go camping together, but this doesn't automatically lead to a harmonious holiday.

"When you go camping on your own you usually don't have any trouble," he says. "You eat whenever and whatever you want. You put your tent up where you like and you sleep whenever you want—everything is just as you want it. But as soon as you go camping with ten people, you have a problem. Everyone starts arguing—why don't we put the tent up here? Why can't we eat now?"

To avoid this kind of problem, cells are constantly communicating with each other, sending and receiving signals that tell them what's going on and whether or not any action is needed. Some of these messages are free-floating chemicals that diffuse from cell to cell or travel around in the bloodstream, while others are more like an old-fashioned "two tin cans and a piece of string" system, requiring direct physical contact between cells for the transmission of signals. In fact, many of the genes required for multicellularity are involved in communication between cells—something that also breaks down during the development of cancer.

While being multicellular has distinct advantages over a unicellular lifestyle, with more cells come more problems. As with so many things in life, the more complicated something is, the more things can go wrong. If you have layers of complex regulatory mechanisms in multicellular organisms that enable cells to specialize into different roles and multiply only when and where they're needed, then you have a lot of possible things that can go wrong and unbalance the whole system. Everything was so much simpler a billion years ago when we were little more than amoebas.

## THROWBACK THURSDAY

In 2011, cosmologists Paul Davies and Charles Lineweaver published a speculative paper entitled *Cancer tumors as Metazoa 1.0: tapping genes of ancient ancestors*, outlining the idea that cancers are an evolutionary throwback (atavism) to an ancient way of life. Rather than being a collection of single selfish cells, Davies and Lineweaver describe cancer as cells that have "devolved" to become a loose collective, similar to the earliest multicellular organisms (metazoans) and running under the control of previously inaccessible ancestral genetic programs. According to their theory, this atavistic behavior emerges as a kind of "safe mode" in response to stressful conditions such as low oxygen levels—exactly the kind of conditions that would have been found on the early Earth around the time multicellular animals were evolving and are also found in the local environment around a tumor.

Few things irritate biologists as much as physicists bounding into their subject area with all the enthusiasm of a puppy with a new toy, proposing overly simplistic solutions that ignore decades of carefully constructed domain knowledge. While their idea that cancer is an evolutionary throwback gathered a fair bit of media coverage, the scientific backlash could have been felt from outer space. Their hypothesis provoked either mockery of the pair's ignorance of cancer biology and genetics, or frustrated bafflement that making banal statements about how the genes involved in the fundamental processes of life are very, very old was apparently enough to earn a glowing reputation in the scientific and popular press for being a "disruptor" or "maverick."*

To a certain extent, Lineweaver and Davies' atavistic theory doesn't hold up to the realities of biology. Instead of being a throwback to a specific point in the history of multicellular life, cancer emerges from cells that have embarked on their own unique evolutionary trajectory within the complex environment of the body. They're subjected to the pressures

* In fact, Davies and Lineweaver weren't the first physicists to come up with the idea that the evolutionary origins of multicellular life might shed light on the problem of cancer. Rafael Sorkin, now professor emeritus at the Perimeter Institute for Theoretical Physics in Ontario, Canada, spent the best part of twenty years trying to persuade a scientific journal to publish his theory that cancer is the result of the breakdown of the control mechanisms that first emerged when unicellular cells began to live together in multicellular bodies. He eventually gave up the struggle and uploaded his thoughts to the online physics repository arXiv in 2000.

of natural selection, utilizing anything they can find in their mutated genomes that enables them to survive (as we'll see in more detail later on). Still, Ruiz-Trillo's work on *Capsaspora* tells us that all that stands between an organized multicellular system and a bunch of autonomous unicellular organisms is tightly controlled gene regulation and cell death. And there are some intriguing emerging observations about what might going on at a genetic level when the multicellular pact between cells starts to break down.

The increasing availability of DNA sequencing data from many different species has enabled researchers to draw up a family tree of life, detailing the relationships between different species and how far back they last shared a common ancestor. This provides a handy way of calculating the age of any given gene. For example, if a particular gene is only present in species on the mammalian branch of the tree, it's a safe bet to assume that it evolved around 65 million years ago when mammals first emerged. But if it's present in everything from bacteria upwards, then it's going to be much, much older and was probably present in our very earliest unicellular ancestors.

A couple of years ago, David Goode—a group leader at the Peter MacCallum Cancer Center in Melbourne, Australia—had the bright idea of mapping the ever expanding repertoire of "cancer genes" emerging from large-scale tumor sequencing projects onto this tree of life. He wanted to see whether or not there was any relationship between the age of a gene and its role in cancer. But it wasn't until a talented young Venezuelan, Anna Trigos, came to his lab in search of a PhD project that he found someone willing to give it a go.

Curiously, Trigos found that the genes that are most active in cancer cells were the oldest. These are the kinds of genes that drive basic functions such as cell proliferation or DNA repair, which date back to the origins of the earliest unicellular life. By contrast, the ones that were least active were the most recent to evolve, mostly specific to mammals or multicellular animals and responsible for more complex jobs, including the creation of specialized organs and communication between cells.

The same pattern holds up across all the types of tumor that she's looked at so far: unicellular genes are activated while more modern multicellular genes are shut down. This suggests that cancer cells are opting out of their regular role in the society of cells and behaving in a more selfish, single-minded manner. It's not so much that they're entirely reverting back to an atavistic, amoeba-like form, but the kinds of mutations that favor the growth of cancer tend to be those that knock out the systems that normally maintain multicellular order and allow cheating cells to prosper.

## YOU CAN'T BEAT THE CHEATS

Cancer is the price of life. Our multicellular bodies are effectively a truce with our cells' inner unicellular tendencies. However, we have to keep some of these singular abilities easily accessible. Rapid cell proliferation is essential to allow stem cells in the blood, bone, bowel, and skin to make the millions upon millions of new cells we need every single day for regeneration and wound healing. Cut out a chunk of your liver and the remaining cells will temporarily reactivate a phenomenal regenerative power that can rebuild two pounds of tissue in a matter of weeks. These processes are incredibly tightly controlled, yet things can—and do—go wrong.

As soon as there's a society of cells, cheaters will arise. This means we can expect to see cancer appearing not only throughout the whole history of human existence, but also across all other multicellular animals, as indeed we do. And if there's complex multicellular alien life out there in the universe, the odds are that most of those creatures will be susceptible to cancer too.

These cheating cells are rebelling against the rules laid down by the genes that construct our tissues and control cell growth. Any alterations in these genes make it more likely that the rules will be bent or broken, allowing tumors to emerge. As we'll see in the next few chapters, understanding how these genetic changes drive cancer has been a key focus of research for more than a century.

But there's something more fundamental going on here. As a writer, I spend my life trying to come up with analogies and metaphors that bring the science of life into sharper focus for people with the interest but not necessarily the expertise to grapple with the finer details of molecular biology. But thinking about the society of cells and its cancerous cheaters, I realize that it isn't a metaphor—that's just how life is. Every multicellular body, group of animals, or population of humans is a society with a set of social contracts and laws, populated by rule-takers and rule-breakers. Cheats will inevitably emerge, particularly once the controls and rules start to get a bit shaky. Inside each one of us is our own personal Judas.

Like a beautiful fractal in which every subunit is merely a smaller version of the whole, this is just how all groups of living things work. From humans to amoebas, queen bees to cancer cells, it's assholes all the way down.

# 3

# YOUR CHEATING CELLS

The question of what causes cancer has occupied us for thousands of years. For much of this time, cancer was seen as a supernatural punishment meted out by deeply offended deities or brought to bear by witchcraft. The Egyptians blamed the wrath of the gods, while ancient Chinese texts put it down to an imbalance of "evil qi." Alongside runs the parallel principle that getting back on the right side of your god of choice through appropriate prayers and rituals should be the path to a cure.* This is a sentiment that still persists today, although orthodox religious explanations are increasingly elbowed out by more nebulous ideas around "wellness." Cancer is seen as

* In nineteenth-century France, religious communities maintained that female masturbation caused tumors in the uterus, while good old-fashioned sex was responsible for cervical cancer. While it's true that sexually transmitted Human Papillomavirus (HPV) infection is a key cause of cervical and other genital cancers (although vastly more people are infected with HPV than ever develop these diseases), we can probably chalk this particular theory up to the patriarchy.

a punishment for living an unhealthy life (or being forced to endure an impure and polluted world), and strict adherence to the rites and rituals of alternative therapy is the only hope for salvation.

Medical minds have long sought more-rational answers. Writing in the fourth century BCE, Hippocrates, the ancient Greek "father of medicine," proposed the idea that the body is made up of four color-coded fluids or humors: red blood, watery phlegm, yellow bile, and black bile. When these vital liquids are in balance, a person is well. If they're out of whack, then the person gets sick, with cancer being caused by an excess of black bile in the system.

The Roman physician Galen took Hippocrates' idea and ran with it, incorporating humor theory into his texts. These became the foundation of clinical practice in Europe as well as in the Islamic world for more than a millennium. Galen also noticed that breast cancer turned up more frequently in women who hadn't had children, leading him to ponder that the disease might be the result of some kind of trapped noxious substance that was expelled through breastfeeding.*

By the 1500s, people were starting to realize that Galen's sense of humors didn't add up, especially when nobody could confirm the existence of the infamous but elusive black bile. A new theory sprang up in the mid sixteenth century based on observations of multiple cases of cancer in members of

* Twentieth-century researchers have shown that while his observation holds true and breastfeeding does have a protective effect against the disease, it's more likely to be working through hormonal differences between women who do or don't lactate.

the same household, leading to the conclusion that the disease must be contagious. Like Galen's hypothesis about toxic breasts, this observation was correct but the cause was wrong: we now know that inherited genetic variations can significantly increase the chances of particular cancers, explaining the high prevalence in certain families. And although some contagious infections, such as the Human Papillomavirus (HPV), can drive cells further down the road toward cancer, it's not normally possible to catch cancer (with some notable exceptions, see page 245). Still, this idea significantly contributed to the fear and stigma around the disease, with early cancer hospitals banished to the outskirts of towns and cities for fear of infecting the populace.

The next big thing was lymph theory, which became popular from the middle of the seventeenth century. This narrowed down Hippocrates' four vital fluids to just two really important ones—blood and lymph—whose correct meanderings around the plumbing of the body were essential for health. One of the main proponents of this concept was the legendary Scottish surgeon and anatomist John Hunter, who believed that tumors arose from pockets of fermented lymph that had leaked out of the bloodstream. Thanks to Hunter and other high-profile supporters, lymph theory survived until the middle of the nineteenth century, when it was shot down by pesky pathologists who used their newfangled microscopes to prove that tumors were actually made of human cells, not congealed fluid.

But although cancers clearly contained cells of some sort, their origins were still a mystery. All sorts of new ideas emerged as scientists struggled to make sense of the weird and wonderful cellular world emerging underneath their microscopes. Some believed that cancers were the result of strange little cells that budded off from a layer of tissue known as the blastema. Others developed their own variations on this theme, imagining that tumor cells spontaneously formed from congealed fluid seeping from blood vessels or were leftovers from our time in the uterus—an idea that actually turned out to be correct in the case of childhood cancer (see page 138). These theories battled it out for the best part of a century, each supported by its own scientific proponents in various parts of Europe and the United States.

Detailed observations eventually revealed that cancer cells adhere to the fundamental principle of biology: *omnis cellula e cellula*—all cells come from cells. Cancer is not a curse, contagion, congestion, or coagulation. It emerges when our own cells cheat on us, proliferating unchecked and eventually spreading to other parts of the body. Still, this only tells us *what* cancer is, it doesn't tell us *why*. What makes a well-behaved cell slip the bonds of good behavior and start doing its own thing?

## SPYING ON CELLS

The nineteenth century was an exciting time to be a man with a microscope (and yes, they were mostly men). The tools and techniques of microscopy were advancing fast, including the development of brightly colored synthetic dyes concocted from the chemicals in coal tar, enabling keen-eyed scientists to spy on the inner workings of cells.

One of these men was German biologist Walther Flemming. He became fascinated by the dark material in the heart of every cell, which he named "chromatin" for the way it soaked up the colored stain. Sifting through preparations of salamander cells, he described how the chromatin rearranged itself into long threads as a cell got ready to divide, with each new cell getting an equal share. Flemming referred to this process as *Karyomitosis*, and the thin threads as *Mitosen*. While the use of the term *mitosis* to describe cell division quickly stuck, it took another German scientist, Heinrich Wilhelm von Waldeyer-Hartz, to come up with the catchy name "chromosomes" to describe these fragile strands of DNA that contain the instructions for life in every cell.

Although virtually nothing was known at the time about DNA or the mechanisms of inheritance, this curious process of chromatin copying and allocation seemed like it ought to be important. In 1890, yet another German, pathologist David von Hansemann, focused his attention—and his microscope—on the strange cells in tumors, noticing that some cancer cells seemed to be going through mitosis in a very weird way. As a normal cell divides, it sets up two "poles" on opposite sides, a bit like the North and South Poles on a globe. The two sets of copied chromosomes are then separated by being pulled to each pole, meaning that each new "daughter" has a complete set of DNA when the cell splits in two across the equator. Rather than having two poles, Hansemann saw that cancer cells often had three or more.

As might be expected, this creates a lot of confusion as a

cell tries to work out how to divide two into three. Furthermore, he noticed that even if cancer cells did manage to get their act together and create just two poles, the chromosomes weren't always equally divided between them. Hansemann suspected that this chromosomal imbalance was a key feature of cancer and quite probably the first step in the process of tumor development. He even went so far as to suggest that cancer cells without obvious mitosis problems might have problems with their chromosomes that were simply too small to see. This was an insight that turned out to be remarkably prescient in the light of the subsequent discovery of genes and the cancer-causing mutations within them.

Unfortunately, Hansemann's ideas failed to catch on—probably because everyone at the time was obsessed with the idea of the budding blastema. As a result, his name is often lost to the history of cancer genetics in favor of the biologist Theodor Boveri (yet another German). In 1914, as Europe was on the brink of immolating itself in war, he published a short monograph titled *Concerning the Origin of Malignant Tumors*, in which he laid out his observations and thoughts on the strange chromosomal dance inside cancer cells. Most of the book is taken up with detailed descriptions of everything that can possibly go wrong during the fertilization of sea urchin eggs—Boveri's organism of choice—including extra sperm, missing chromosomes, multiple poles, and paralyzed mitosis, any of which spelled bad news for the urchin's chances of life.

Even though Boveri barely did any observations on cancer cells, he was intrigued by the connections between the chromosomal chaos in his urchin eggs and the unusual activities of cancer. He suggested that tumor cells were driven to run amok by a "specific abnormal chromosome constitution" (with a nod to Hansemann having come up with the idea first). Boveri also came up with the idea that there might be "inhibitory chromosomes," which, if lost, would enable a cell to start proliferating uncontrollably, neatly foreshadowing the discovery of tumor suppressor genes. He also had the insight (or possibly made a lucky guess) that the chromosomal units responsible for particular traits—which we now know to be genes—are likely to be lined up along each of the chromosomes in a specific order. Boveri died in 1915, just a year after publishing his little book, leaving his American wife, Marcella, to translate it into English and bring his ideas to a wider audience.

The idea that cancer was the result of normal cells suffering from an internal chromosomal malady grew in popularity through the early decades of the twentieth century. American pathologist Ernest Tyzzer first used the phrase "somatic mutation" in 1916 to describe the idea that tumors arise owing to some kind of modification or alteration of hereditary material in the regular (somatic) tissues of the body. By 1922, the eminent fruit fly geneticist Thomas Hunt Morgan had pinned the blame on faulty genes, proposing that "cancer is due to a recurrent somatic mutation of a specific gene . . . that leads to cancer." And thus, the "somatic mutation theory of cancer" was born: the idea that cancer is caused by mutations in genes in normal cells that cause them to go awry and proliferate out of control.

Throughout the twentieth century, scientists started to

piece together how it all worked. We now know that chromosomes are long strands of DNA made from four chemical building blocks (bases) that are strung together in endlessly varied combinations. Genes are shorter stretches of DNA scattered within each chromosome. They're the biological instructions that tell our cells when to grow and multiply, what job to do in the body and even when to die. It's the order of these bases—adenine (A), thymine (T), guanine (G) and cytosine (C)—that convey the information within a gene, effectively acting like a molecular alphabet spelling out the recipes of life.

The human genome—that's the complete set of DNA required to make a person—contains 20,000 genes, give or take a few thousand,* distributed across twenty-three pairs of chromosomes. However, actual genes make up less than 2 percent of the genome. The rest is often referred to as "junk DNA" in the popular science media (the more technically correct term being "non-coding DNA," because it doesn't contain the coded genetic instructions for the production of protein molecules). Some of it contains the million or so control switches that turn genes on and off, along with structural components that are necessary for maintaining the right number and length of chromosomes as cells divide. There are stretches that spin out strings of molecular thread known as

* It may seem surprising in this age of high-tech DNA sequencing, but the precise number of genes in the human genome is a contentious issue, as it depends on your definition of exactly what constitutes a gene in the first place. There's much more about this in my first book, *Herding Hemingway's Cats*.

non-coding RNA, which either perform important functions in their own right or help to control the activity of other genes. And there's also plenty of genetic junk that seems to be genuinely useless, although exactly how much is a hotly debated topic in genetics.

Thousands of genes have to be switched on in exactly the right place and at the right time as we grow from a single fertilized egg cell to a baby and then to an adult, along with just the right amount of cell proliferation to build and maintain a body. So it makes sense that any changes to the "letters" of this genetic blueprint will cause problems. A spelling mistake in an important gene might cause a cell to start multiplying out of control. Alternatively, it might damage the instructions that tell cells to die when they're damaged or worn out. Further alterations in other vital genes lead to an aggressive, unstoppable tumor, eventually gifting some of these rogue cells with the ability to break free and start wandering around the body.

This line of thinking has led to the idea that cancer is fundamentally a disease of DNA, with two logical conclusions. First, if we can figure out exactly which faulty genes and molecules are driving the unchecked growth of cells in a tumor, then we should be able to find targeted "magic bullets" that stop them (an idea we'll come to later). Secondly, it suggests that we should be able to work out exactly what causes cancer: if you can detect the DNA changes (mutations) that have led to the development of an individual person's cancer and work out what caused them, then you should be able to figure out the solution to their biological whodunnit.

Black bile and angry gods aside, probably the oldest

documented scientific study linking a type of cancer with a specific cause is the treatise *Cautions Against the Immoderate Use of Snuff*, published in 1761 by British physician and botanist John Hill. Over more than fifty gruesome pages, Hill describes the growth of hard tumors inside the nostrils of men who sniffed tobacco, hoping that his words will be "useful to such, as from an excess in the same practice, may be liable to like disorders."

The even lengthier work *Chirurgical observations relative to the cataract, the polypus of the nose and cancer of the scrotum* was published fourteen years later by Percivall Pott, an eighteenth-century British surgeon with a keen professional interest in the nether regions of London's underage chimney sweeps. Most of these young boys were abandoned, abused and unwashed, sent up chimneys naked or clad in loose trousers and shirts. A significant number went on to develop painfully destructive genital tumors known as soot warts.

Not only did Pott find that these cancers could be cured if they were operated on quickly before the disease started to spread, but he also realized that the cause must lie in the accumulation of soot on their unwashed testicles. His findings led him to campaign for improvements in working conditions, frequent baths and the provision of the kind of tight-fitting protective clothes that were worn by sweeps in Germany and elsewhere. Across Europe, these public health recommendations led to the virtual eradication of the cancers within a matter of decades. But back in Britain, Pott's efforts to bring about better conditions were thwarted by the concerns of wealthy homeowners, insurance companies and gangmasters whose income depended on the boys' slave labor. Thanks to their arguments that it was better to sacrifice a few poor kids than risk pollution and chimney fires for the rich, the horrendous conditions—and the associated horrendous cancers—persisted well into the nineteenth century.

It wasn't until the 1930s that scientists could confirm that Pott's hypothesis was correct, by showing that painting solutions of soot onto the shaved skin of mice would induce tumors to form. This method revealed the carcinogenic potential of a number of nasty chemicals, including benzpyrene and polycyclic aromatic hydrocarbons (PAHs). Some of these are also found in tobacco smoke, providing the first inkling that the hugely popular habit—glamorized by film stars and endorsed by doctors—might not be that good for you after all. Coming from the other direction, by 1950 British researchers Richard Doll and Austin Bradford Hill had published a study of more than 2,000 patients admitted to twenty London hospitals, showing that smokers were much more likely to be diagnosed with lung cancer compared with non-smokers or people with other types of cancer.

Although Doll and Bradford Hill are widely credited with being the people who proved the link between lung cancer and smoking, the connection had actually been made more than a decade earlier. However, there was one significant reason why this earlier work demonstrating dangers of tobacco was ignored: it was done by Nazis.

Scientists working at Nazi-controlled Jena University in Germany during the 1930s were the first to show a link between smoking and cancer in humans, even coining the term "passive smoking." Yet, because these findings were published in German in the midst of World War II, very little attention was paid to them. Perhaps more significantly, Jena was an academic hotbed for the unethical and deeply flawed racial science that helped to form the basis of the Nazi's eugenic policies that led to the deaths of millions of "undesirables," including Jews, Roma gypsies, gays, ethnic minorities, and disabled children and adults. So there was understandable skepticism about any of the research emerging from the university at that time, regardless of whether or not it was scientifically accurate. And Hitler himself was a staunch anti-smoking fanatic, which probably didn't do much to help the cause of tobacco control.

More non-Nazi evidence came in 1948 in a paper by cancer surgeon Willem Wassink at the Antoni van Leeuwenhoek Hospital in Amsterdam, Netherlands, showing that heavy smokers were twelve times more likely to have lung cancer than non-smokers. But because it was published in Dutch, this knowledge didn't make it into English language medical textbooks. Even earlier, Argentine doctor Ángel Roffo discovered in 1931 that tobacco tar extracts could cause cancer when painted onto the ears of rabbits while pure nicotine didn't. But because he published his results in German journals, which was where all the hot smoking research was happening at the time, they went unnoticed in the English-speaking world.

Regardless of the tendency of the medical establishment to ignore anything done by Nazis and/or not published in English, even the British pairing of Doll and Bradford Hill struggled to get people to take notice of their findings, which they considered to be cast-iron proof of the role of smoking in lung cancer. Doll's membership of the Communist Party was seen as reason enough to ignore his work, even though Bradford Hill was firmly embedded within the establishment. The only way that they could get doctors to sit up and take notice was to show that not only was smoking killing their patients, it was killing them too.

In 1951, they recruited 40,000 doctors into the most ambitious research study ever done at the time, following them up over years to unravel the connections between smoking and health. They didn't even have to wait that long to get their first answers. By 1954, the figures clearly showed that lung cancer was twenty times more common in smokers than non-smokers. And by 1956, solid links had been found with a host of other ailments including heart attacks, chronic lung disease, and esophageal cancer. Even so, it wasn't until the 1960s that restricting tobacco sales and marketing for the sake of the public's health was seen as a good idea, and sales only started to fall in the early 1970s.

In the modern era, our answers to the question of what causes cancer are somewhat more informed. A quick search online reveals a plethora of ideas from the obvious to the obscure. Smoking, bad diets, nasty chemicals in the environment, UV rays from the sun, certain viruses, inherited gene faults, pollution, a faltering immune system . . . the list goes on. Then there's the weirder suspects that turn up in the media from time to time, with some of my favorites including shower curtains, turning the light on at night to go to the bathroom, and—most ridiculously—water. The Daily Mail

Oncology Ontology project was a short-lived blog attempting to catalog the newspaper's love of dividing all inanimate objects into things that either cause or cure cancer, although the anonymous writer quickly gave up after realizing the sheer scale of the task.

But, as any good statistician knows, correlation doesn't prove causation: merely finding that many people with the same type of cancer have been exposed to the same thing doesn't necessarily prove that it's to blame. Cause is a slippery word when it comes to cancer, as it implies that just one thing triggered the disease. There are plenty of smokers who don't get cancer during their lifetime and plenty of non-smokers who do, yet the evidence overwhelmingly shows that people who smoke have an increased chance of developing the disease compared with those who don't.* So it's better to think about these kinds of things as *a* cause rather than *the* cause, or refer to them as "risk factors" if you're feeling technical.

We can say that something is a cause of cancer if exposure to it increases the probability of getting cancer, all other things being equal among similar people of the same age. By way of analogy, if you took 2 million drivers, randomly allocated half of them to drink four double whiskies, and sent them all on a

* While I was working at Cancer Research UK, I lost count of the number of stories I heard along the lines of, "Well, my grandfather smoked all his life and never got cancer . . ." Funnily enough, my response of a brief lecture about statistics and risk factors was usually unappreciated, while telling them, "Well, my granny smoked and she did" tended to shut them up.

journey down I-45 from Dallas to Houston, not only would you be well advised to steer clear of the highway, but you'd also expect more accidents to happen in the group who'd had the Scotch. But you'd also expect many of the drunk drivers to make it to the capital unscathed, and you wouldn't be surprised if even a few of the sober ones had an accident along the way for an unrelated reason.

Still, such talk of risks and probabilities doesn't satisfy molecular biologists keen to know exactly *how* these things cause cancer. If, as the somatic mutation theory suggests, cancer is caused by faults in specific genes within cells, then we should be able to search for the fingerprints of damage left within the genome by cancer-causing agents. And to do that, we need to be able to read the recipe book of life.

## READING THE RECIPES

The first reliable method for reading the order of molecular "letters" (bases) in a string of DNA—a technique known as DNA sequencing—was invented in the late 1970s by British biochemist Fred Sanger, whose name now adorns the Wellcome Sanger Institute in Cambridge, one of the largest DNA sequencing facilities in the world. Sanger's original technique was time-consuming and cumbersome, allowing scientists to read a couple of hundred bases at best. So rather than looking at all 6 billion letters of the human genome in search of cancer-causing mutations, researchers started by focusing on just one gene, *TP53*, which encodes the potent protector protein p53. This powerful molecule normally suppresses the development of cancer by making cells die if their DNA is damaged beyond repair, and faults in the gene or its control mechanisms are present in the majority of human tumors.

By the 1990s, researchers in the United States had managed to show that many different types of cancer had their own unique suite of mutations in *TP53*. Each of these alterations was likely to have been caused by different agents, with some of them—such as the chemicals in tobacco smoke or ultraviolet (UV) light from the sun—leaving distinctive patterns of DNA damage behind.

The idea that certain carcinogens would leave characteristic marks within the genome of cancer cells intrigued Mike Stratton, then a young geneticist hunting for mutations in cancers affecting the muscles and other soft tissues, and now Director of the Wellcome Sanger Institute. If these cancer-causing culprits leave their fingerprints in *TP53*, he wondered, what about the other 19,999 or so genes in the genome? Or all the rest of our DNA? Frustratingly, the technology of the time wasn't up to the task of scrolling through the billions of bases in the human genome, so he had to be patient.

The solution arrived fifteen years later in the form of next-generation sequencing: DNA-reading machines enabling scientists to move from reading hundreds of bases at a time to thousands or even millions. Straight away, Stratton saw the potential for the technology to revolutionize our understanding of the genetic changes inside individual tumors, setting the Sanger Institute's huge banks of DNA sequencing machines in motion to read every letter of DNA in a single tumor.

By 2010, he and his team had produced the first whole cancer genomes. These were detailed maps showing all the genetic changes and mutations that had occurred within two individual cancers—a melanoma skin cancer and a lung tumor from a smoker. These choices were far from accidental: decades of population studies and lab research had shown that UV light is likely to be the strongest risk factor for melanoma, while the link between tobacco and lung cancer is well known, with cigarette smoke containing more than sixty carcinogenic chemicals.

With such strong lead suspects, Stratton and his team had the best chance of finding clear fingerprints in the genome. But although they expected to see the same kinds of mutations that had been picked up in the original *TP53* studies, they weren't prepared for the sheer scale of genomic vandalism that they uncovered.

The lung cancer was shot through with nearly 23,000 mutations, with 132 of them hitting genes. There were hundreds of small sections that were missing or duplicated, and more than fifty large-scale rearrangements, with whole bits of one chromosome cut and pasted elsewhere. And, as might be expected, the genome was plastered with characteristic fingerprints of tobacco-related damage. The melanoma was even worse: more than 33,000 single "typos," many of which bore the classic hallmarks of UV damage, along with extensive chromosomal cutting, pasting, and reorganization. The sheer scale of mutations eclipsed anything that had been measured before in any human cancer and the results were a powerful proof that the fingerprints of specific carcinogens could be seen in the genome.

Stratton and his team started to expand their search, looking at the patterns of mutations in other tumor types. But there was a problem: with thousands upon thousands of mutations in a typical tumor, the detective work becomes a lot trickier for cancers without such an obvious prime suspect as those first lung and skin tumors. And even those two, which are each conventionally thought to have one major risk factor, contained many mutational fingerprints that didn't match the destructive work of tobacco or UV.

Making sense of the mess of mutations in a tumor genome is a bit like being a forensic scientist dusting for fingerprints at a crime scene. You might strike it lucky and find a set of perfect prints on a windowpane or door handle that match a known killer in your database. But you're much more likely to uncover a mishmash of fingerprints from a whole range of folk—from the victim and the murderer to innocent parties and police investigators—all laid on top of each other on all sorts of surfaces. So how do you work out whose prints are whose? And then how do you figure out whodunnit?

Fortunately, Stratton's PhD student Ludmil Alexandrov (now an assistant professor at the University of California, San Diego) came up with a way of solving the problem. He realized that the individual mutational fingerprints in a tumor can be distinguished using a mathematical method called blind source separation, previously used to separate data from multiple sources; for example, splitting out individual vocal and instrumental tracks from a single audio file.

Alexandrov's algorithm unearthed twenty distinct mutational signatures from nearly 5 million mutations in more than 7,000 tumors, covering thirty of the most common forms of cancer. Some fingerprints turned up in every single tumor, while others were specific to just a handful of cancer types. All of the cancers had at least two different signatures, while some had at least six. A few years later, that number had risen to at least thirty unique mutational fingerprints, each caused by a different agent. An even larger analysis of nearly 85 million mutations in around 25,000 cancers has now raised the number to around sixty-five, although probably only around fifty of these are truly unique.

We're now starting to understand how many of these characteristic patterns arise. Carcinogenic chemicals cause mutations by physically binding to specific bases and affecting their shape. These alterations throw a molecular wrench in the works, holding up fundamental processes such as copying DNA or reading genes, so they have to be fixed to keep the cell healthy and functioning properly. For example, benzopyrene (one of the major carcinogens in tobacco smoke) tends to bind to G bases, as does aflatoxin, a cancer-causing chemical made by certain molds. But each of these types of damage is repaired in a specific way, leaving a characteristic change in the DNA sequence.

By contrast, UV light leads to mutations by causing neighboring Cs to fuse together. This unusual shape is interpreted as a pair of Ts by the DNA-copying machinery, resulting in a permanent change in the DNA sequence at that position. Aristolochic acid—a chemical found in plants of the *Aristolochia* (birthwort) family*—leaves a different mark, with pairs of AT bases flipped round to TA. Curiously, benzopyrene only leaves its damaging fingerprints in smoking-related cancers of the lung and larynx—tissues that are directly exposed to smoke. But we know from large-scale population studies that smoking increases the risk of several other tumor types, including bladder, pancreas and kidney cancers, so there must be another mechanism at work. The Sanger team have also identified a mysterious new mutational signature in all smoking-related cancers, distinct from the marks left by benzopyrene. This may be a possible second culprit for causing smoking-related cancers, although its identity is currently unknown.

While we might want to focus on external causes of cancer—particularly nasty chemicals, smoking, or radiation—many of the mutational processes that leave their mark in the genome are the biological equivalent of an "inside job." The mechanisms of life aren't perfect at the best of times and every time a cell repairs or copies its DNA there's a risk that it might accidentally end up making a mistake. The chances of this happening increase dramatically if

* Aristolochic acid occasionally turns up in traditional Chinese herbal remedies, or as a contaminant in wheat produced from fields where *Aristolochia* plants grow as weeds. Its role in causing cancer became starkly clear when more than 100 people in Belgium were admitted to hospital with severe kidney damage, with many of them going on to develop kidney and other urinary cancers. It turned out that all of them had taken herbal remedies or slimming supplements made from plants with high levels of the chemical.

there's any problem with the repair or copying process, such as an inherited or randomly occurring mutation in a component of the molecular machinery.

Each activity leaves an identifiable fingerprint behind. For example, there's "the clock" mutation—C changed to T—caused by the incorrect repair of 5-methylcytosine, a molecular "tag" that's added to DNA as part of normal genetic regulation. This pattern can be used as a biological readout of age, because these particular mutations gradually accumulate as the years tick by. Another obvious fingerprint can be pinned on DNA repair problems resulting from inherited faults in the *BRCA1* or *BRCA2* "breast cancer" genes, often turning up in breast, pancreatic, and ovarian cancers.

A signature caused by problems with a type of DNA mending known as mismatch repair turns up in seventeen cancer types, most commonly bowel and uterine cancers. And many blood cancers contain the fingerprints left behind by errors in the genetic cutting and pasting process that generates the protective antibodies required by the immune system. Misbehavior of a specific DNA copying enzyme known as POLE peppers the genome with a huge number of distinctive mutations and has been found in at least six types of cancer. And then there are mysterious "mutators" known as APOBEC proteins, which usually protect our cells from infection by cutting up DNA from invading viruses. Mike Stratton and his team have found APOBEC fingerprints in more than twenty types of cancer, especially cervical and bladder tumors, yet they have no idea what caused it to turn its power on human rather than viral DNA.

Even so, the culprits responsible for around half the mutational fingerprints found in cancer genomes are still at large. And, even more confusingly, some things that we know cause cancer from large-scale population studies or animal experiments appear to leave no trace in a tumor's DNA. However, the fingerprint database is growing fast. In 2019, Professor Serena Nik-Zainal and her team at the MRC Cancer Unit at the University of Cambridge published the results of an epic five-year study looking for the mutational signatures left by nearly eighty potential carcinogens in the undamaged genomes of healthy human stem cells grown in the lab.

As well as picking a bunch of known genomic criminals like benzopyrene, sunlight, and aristolochic acid, Nik-Zainal's team also looked at the impact of common cancer drugs and gamma rays, along with a bunch of nasty-sounding chemicals. Some of them were known to cause cancer in humans, while others were highly suspicious based on evidence from animal or lab studies. They found the same signatures from the known carcinogens that Mike Stratton and his Sanger colleagues had found in tumor samples, confirming that their technique worked, as well as some new ones. In total they found fifty-five distinctive DNA fingerprints from these potential carcinogens, which can now be matched with signatures that turn up in cancers. Intriguingly, they also found a specific mark that seems to be related to the stressful conditions of being grown in a plastic dish and being doused in chemicals, rather than being a genuine signature of cancer-causing DNA damage "in the wild."

Even as scientists tease out the fingerprints left behind by all kinds of mutagens, being able to write a list of all the perpetrators of damage in an individual person's cancer still leaves us with the same problem of risks rather than causes. In a plot twist worthy of Agatha Christie's *Murder on the Orient Express*, it's becoming clear that we're not dealing with individual "baddies" but a gang of miscreants, all of whom administer a potentially fatal blow to the genome. Each causes mayhem in its own way, but they can combine together to bring about a catastrophic outcome. Even worse, much of this damage is done by the normal hurly-burly of life within our cells, which we can't avoid. But activities like smoking and excessive sunbathing are like a gangster bringing a massive bag of guns into an already edgy situation, significantly increasing the chances that something bad is going to go down.

This means that it's almost impossible to pinpoint one specific cause of a given tumor or say exactly what caused it. A cell may be riddled with mutations, accumulated from all sorts of processes over a lifetime, but if none of them hits the vital genes or control switches responsible for governing growth or death, then it will remain healthy. And because every tumor is shot through with many thousands of mutations, it's impossible to say which culprit delivered the final blow that ultimately tips an unhappy cell into becoming a cancer. And—as we'll see later—there's more to a cancer than just mutations. Even so, we're starting to build a much better picture of the contributions of different risk factors—be they inherited, biological, or environmental—to each individual cancer.

It's taken more than a century to get here, but the development of the somatic mutation theory of cancer—together with the discovery of the extent of damage in the genomes of tumors—has given us deep insights into the causes of cancer. However, while all this work was going on to pin cancer on chemical causes, there was a completely different strand of research running in almost entirely oblivious parallel.

## FROM JACKALOPES TO CANCER GENES

One day in 1932, Douglas Herrick and his brother Ralph went out into the woods near their home in small-town Wyoming to catch a jackalope. These mysterious horned rabbits have existed in folklore around the world for hundreds of years and are the stuff of cowboy legend in the American Midwest.* So when the Herrick brothers turned up at a local hotel with a stuffed, mounted specimen, the owner jumped at the chance to buy it. From that single sale, the Herricks sparked a thriving local industry flogging anything and everything to do with this "half bunny, half antelope, and 100 percent tourist trap," as Douglas' obituary in the *New York Times* put it, sealing the jackalope's fate as a Wyoming icon. Of course, their original jackalope was a piece of taxidermy trickery, stitched together from the skin of a rabbit and the antlers of a deer, as were all the ones that followed. But there's an even stranger biological truth behind the legend.

Just one year after the Herrick brothers began their jackalope-stuffing business, an American virologist named

* Legend has it that jackalopes are fond of campfire singalongs, apparently joining in with a dulcet tenor.

Richard Shope managed to get hold of a cottontail rabbit with large horny growths protruding from its face—exactly the kind of thing that might be mistaken for antlers by a confused cowboy. Shope ground up the horns and filtered them through a sieve so fine that only the very tiniest particles could get through, then applied the resulting liquid to the skin of an unaffected bunny. Sure enough, that animal turned into a "jackalope" too, proving that the condition must be caused by an infectious virus—the only thing small enough to make it through Shope's sieve.

The (real) jackalope's horns are caused by an agent now known as Shope papillomavirus and they're a striking demonstration of the power of viruses to cause strange, unruly growths in the body, including cancer. The first cancer-related virus was found in 1911 by American virologist Francis Peyton Rous, who discovered that "infectious particles" (now known as Rous sarcoma virus) were responsible for spreading a type of soft-tissue cancer known as sarcoma between chickens. Since then, a plethora of cancer-causing viruses has turned up in animals, including poultry, cats, mice, sheep, dogs, cows, reptiles, and even fish, causing leukemia, breast tumors, lung cancers, and more.

The idea that viruses can cause tumors sparked a wave of research through the first half of the twentieth century. There was a huge amount of excitement that viruses would be the explanation for all cancers, along with the tantalizing hope that they might be preventable with simple vaccines. The cover of *Life* magazine in June 1962 boasted the strapline "New Evidence That CANCER MAY BE INFECTIOUS" alongside a picture of Marilyn Monroe, in bigger typeface than the name of the iconic actress. But, as often turns out to be the case, things weren't quite so simple as we might have hoped.

Over the years, scientists have unearthed several viruses that are associated with cancer in humans, the most famous being human papillomavirus (HPV), which is responsible for cervical cancer, as well as other cancers affecting the genitals and anus, in addition to the mouth and throat (possibly transmitted by oral sex). Frustratingly, while most of the animal viruses seem to cause cancer directly—if an animal is infected, it's likely to get the disease—human cancer viruses are more subtle and confusing. If you're sexually active, you're probably going to be infected with HPV at some point in your life. But only a tiny proportion of people with HPV go on to get cancer, while most of us get rid of the infection without a problem.

Another viral culprit is Epstein–Barr virus (EBV), which is linked with Burkitt's lymphoma, and tumors in the nose and throat (nasopharyngeal cancer). Intriguingly, although EBV infection crops up all over the world, the associated lymphoma is mostly found in certain parts of Africa, while EBV-related nasopharyngeal cancer is common in southern China. This suggests there are specific environmental and genetic factors at play, with malaria infection fingered as a likely co-conspirator in Africa. There's a virus that causes T-cell leukemia, hepatitis B and C are risk factors for liver cancer, and human herpesvirus 8 can trigger Kaposi's sarcoma in people whose immune systems are already suppressed by HIV. The most recent addition to the line-up is Merkel cell polyomavirus, discovered in

2008, which affects oval-shaped cells in the skin responsible for detecting light touch sensation.

Altogether, viruses are now known to be responsible for at least a tenth of all cancers worldwide—a staggering 2 million cases every year. Unfortunately, most of these cases happen in poorer countries, making them much less interesting to pharmaceutical companies with their eyes on the bottom line. Despite a growing roster of known tumor viruses, the research community mainly lost interest when it became clear that viruses wouldn't lead to the universal explanation or cure that they'd hoped for. But all that work wasn't a total waste of time. Importantly, the identification of cancer viruses laid the foundation for the discovery of cancer genes. And everybody *loves* those.

# 4

# FIND ALL THE GENES

One of the perks of my job is getting to meet legendary scientists in their natural habitat. I've visited hundreds of labs and offices, seeking out a sense of who someone is and what drives them. Some researchers are hoarders, piling every available surface with stacks of papers so high that misadventurous PhD students are probably still lost in there. Others prefer a more minimalist approach—one Nobel laureate I visited had virtually nothing in his office except a coconut adorned with a pair of googly eyes and a top hat. Visiting the office of pioneering cancer researcher Robert Weinberg at the Whitehead Institute for Biomedical Research in Boston, I'm transported into a whole different world.

At least two walls are plastered from floor to ceiling with photographs of family, lab mates, faculty, and friends going back decades, some still bright and fresh and others faded to a pale sepia wash. In sharp contrast, the large windows are almost completely obscured by a small jungle of houseplants.

I scan nosily through the pictures while he's away fixing us some drinks, hoping to spot famous faces from the past half-century of molecular biology, then quickly duck back into my seat when he returns.

His language is precise, somewhat technical, perfectly grammatically constructed, and sounds like the words of someone who has spent a very, very long time thinking about all of this stuff. Born to Jewish parents who fled from Nazi Germany, the young Bob Weinberg started his academic career studying medicine at the Massachusetts Institute of Technology (MIT), quickly switching to molecular biology when he discovered that doctors were expected to stay up all night tending to their patients. His interest was fueled by the explosion of discoveries throughout the mid sixties, with a fast-paced rush of new tools and techniques revealing the secrets of the genetic code and the underlying molecular mechanisms of life. *

The so-called "central dogma" of biology had been clear for more than a decade: DNA contains genes, which are copied into a related molecule called RNA. This is then "read" by the machinery inside cells to make proteins. Yet the biological underpinnings of cancer were still very much a mystery at the time. Although it was clear by that point that cancer was driven by genetic changes inside cells (the somatic mutation theory we met earlier), it wasn't clear how these alterations made them run out of control. From tumor viruses to

* There's much more about the race to crack the genetic code and other major advances during this period in Matthew Cobb's excellent book, *Life's Greatest Secret* (Profile Books, 2015).

carcinogenic chemicals to strange chromosomal abnormalities in cancer cells, scientists were struggling to figure out how to bring together this seemingly disparate jigsaw puzzle of observations into one coherent solution.

The first clue to what the bigger picture might look like came from cancer-causing viruses, which are little more than a handful of genes wrapped in a protein coat. Despite it being nearly sixty years since Peyton Rous first discovered his eponymous sarcoma virus in an unlucky chicken, it wasn't until 1970 that researchers discovered how it worked.

By comparing two versions of the virus, one of which made chicken cells proliferate uncontrollably and one that couldn't, American molecular biologists Peter Duesberg at the University of California, Berkeley and Peter Vogt at the University of Washington in Seattle discovered that a crucial alteration in just one gene was enough to make all the difference. They called this gene *v-Src* (pronounced "sarc"), with the "v" denoting that it was viral in origin and the "src" signifying sarcoma. These viral "cancer genes" quickly became known as oncogenes, derived from the Greek word for tumor, *onkos*, from which we also get oncology—the study of cancer.

The discovery of *v-Src* made perfect sense in terms of the viral theory of cancer. When the virus infects a cell, its *v-Src* somehow hijacks the normal control mechanisms, causing wayward proliferation that eventually results in a tumor. So it came as a big shock a few years later when researchers found that normal, healthy chicken cells also appeared to have a version of the *Src* oncogene lurking within their DNA. Yet this gene was most certainly avian, rather than viral in origin.

Further studies revealed more of these so-called proto-oncogenes within the genomes of fish, mice, cows, and humans, suggesting that they were regular animal genes that had somehow got picked up by viruses and taken along for the ride at some point in history. The deeper implication was mind-blowing: maybe most cancers weren't caused by viruses at all but were actually driven by corrupted versions of the normal instructions for life.

These revelations gave Weinberg an idea. If carcinogens work by damaging normal genes and turning them into cancer drivers, then these proto-oncogenes must be responsible for the fundamental processes of cell growth and division. All he had to do was find them.

His solution was to extract the DNA from cells that had been treated with cancer-causing chemicals, chop it up into little bits, then stick each fragment into a healthy cell and see if any of them grew out of control into little clumps—a process known as "transformation." Then all he had to do was try to fish that piece of DNA back out of its new host and uncover the identity of the rogue mutated gene (or genes) located within. It might sound simple, but with the tools available at the time it was a long and arduous task that fell to Weinberg's graduate student, Chiaho Shih (now a leading scientist in his own right at the Academia Sinica in Taiwan).

The first surprise was that it worked at all. Against all expectation, Shih managed to take DNA from these chemically mutated cancer cells, put it into normal mouse cells, and make those healthy cells turn cancerous. This proved beyond all doubt that cancer was a genetic disease and could be caused by chemical damage to the cell's own genes, rather than purely being the result of an external infection.

The next challenge was to work out the identity of this broken genetic culprit. Unfortunately, it was virtually impossible to distinguish the chemically altered gene from the rest of the DNA, as both the host cells and the extra genes all came from the same species. To solve this problem, Weinberg and Shih turned their attention to human bladder cancer cells that were growing happily in Petri dishes in the lab.

Using the same approach as before, Shih smashed the human cancer DNA into tiny pieces and transferred each one into mouse cells. Pleasingly, he saw the same effect: there was one fragment of tumor DNA that made the mouse cells start growing like crazy, which must contain an oncogene. By exploiting the differences between mouse and human DNA, Weinberg and Shih were able to extract this elusive cancer-causing gene and show that it matched a sequence in the normal human genome. Shortly after they'd done all this work, Weinberg realized that their mystery gene was almost exactly the same as *v-Ras*, a gene that had already been identified in a rodent tumor virus—knowledge that might have saved two years of hard work and heartache along the way.

Like *Src* before it, *Ras* was a normal cellular gene with the power to drive cells out of control when mutated. In common with many other oncogenes that have been discovered since then, *Ras* is a type of molecular "switch" known as a kinase. Mutations in *Ras* create a switch that's permanently flipped to "on," constantly sending signals telling the cell that it's time to divide. We now know that around one in five cancers in many different organs have a mutation in *Ras*, highlighting its critical role in the journey from healthy cell to tumor. The work of Weinberg and many others turned the 1980s and 1990s into the era of Finding All the Genes, with researchers turning up more and more oncogenes, many of them with counterparts in cancer-causing viruses.

Another source of cancer genes came from studying whole chromosomes in tumor cells, echoing back to the work of Boveri and Hansemann at the turn of the century. Since those days, keen-eyed microscopists had been developing ever more clever ways of studying chromosomes, although it wasn't until the 1950s that techniques were reliable enough to prove that humans have forty-six chromosomes (twenty-three pairs) in every cell.*

In 1959, two researchers at what's now the Fox Chase Cancer Center in Philadelphia, Pennsylvania—David Hungerford and Peter Nowell—made a very strange discovery in the cells of a patient with chronic myeloid leukemia (CML), a type of blood cancer. Although all the chromosomes were present and correct, one of the two copies of chromosome 22 was unusually small. Already a puny entry in the line-up of human chromosomes, this abnormal chromosome 22 was less than half the size it should have been.

Nowell and Hungerford looked at cancer cells from other

* For a long time the accepted number was forty-eight (twenty-four pairs), with the correct answer only being determined in 1955 by Indonesian-born geneticist Joe Hin Tjio. I find it amazing that people were unsure as to how many chromosomes made up the human genome as late as the mid-1950s.

people with CML and there it was again—a stubby stump where a full-sized chromosome should be. Publishing their findings in a brief 300-word paper in 1960, their discovery launched a scientific odyssey that would set the course for cancer drug discovery for decades to come.

It took another decade or more to figure out how this chromosomal change is related to the development of leukemia. The diminutive Philadelphia chromosome, as it became known, results from a chromosome fusion—a genetic cut-and-paste event in which a tiny part of a larger chromosome, number 9, gets switched for the bulk of the already petite 22. Crucially, it brings two genes together that should never normally touch—*BCR*, a gene whose function is still unknown, and *ABL*, a powerful oncogene that was first discovered in a mouse leukemia virus.

This fusion creates a monster. The hybrid *BCR-ABL* gene encodes a permanently hyperactive kinase that constantly drives proliferation, endlessly spinning out new cells into the blood. The search for drugs that could stop this overexcited mutant molecule in its tracks led to the development of Glivec (imatinib), arguably one of the most successful cancer drugs of all time. It has transformed survival from CML since it came on the market in 2001, earning billions of dollars for its parent company Novartis in the process.* Just before the introduction of Glivec, around four in ten people with

* The full story of the science and history behind the development of Glivec is brilliantly told by Jessica Wapner in her book *The Philadelphia Chromosome: A Mutant Gene and the Quest to Cure Cancer at the Genetic Level* (The Experiment, 2013).

CML could expect to survive for at least five years after diagnosis. Nearly two decades later, around seven in ten make it to five years. And someone with CML whose cancer has disappeared after just two years of treatment with the drug is now likely to live as long as someone who has never had the disease.

Many other research groups were keen to follow in the footsteps of the Philadelphia discovery in the hope of finding more fusion genes responsible for driving cancer, especially if they were as amenable to lucrative drug development as *BCR-ABL*. As technology improved, researchers started making detailed maps based on the alternating patterns of light and dark bands within chromosomes, revealed with a purple dye known as Giemsa stain. This sticks to certain sequences in DNA better than others, creating characteristic bar code-like stripes that distinguish each chromosome and reveal any large-scale alterations.

Next came vivid chromosome painting, using clever combinations of multiple fluorescent dyes to daub each chromosome a different color and show how the genetic deck had been reshuffled in cancer cells. DNA sequencing also proved a useful tool for finding fusions as it became cheaper and faster. But despite the discovery of fusion genes in many cancers, none of them have been able to repeat Glivec's stellar success.

The discovery that genetic mutations drive tumor growth by triggering unwanted cell proliferation is just one side of the coin. Cancer isn't just a disease characterized by excessive life, it's also a problem of not enough death. Our bodies are churning out millions of new cells every single minute of the day—blood and bowel, skin and bone, and everything in between—all of them born to die. This constant cycle of self-renewal is a powerful cancer protection mechanism: if a cell is dead, it can't multiply out of control.

There are important fail-safes on top of this, such as molecular toolkits that press pause on the cell cycle so DNA damage can be fixed. And there are cell suicide pathways that get activated if things are beyond repair, controlled by genes known as tumor suppressors. If oncogenes act like a gas pedal, pushing full speed ahead into the cell cycle, then tumor suppressors are the brakes, holding things back until it's safe to proceed. For a cancer to develop, not only does the accelerator need to be jammed to the floor, but the brake cables have to be cut, too.

## FROM BROCA TO BRCA

As you're reading these words, you're using part of your brain called Broca's area. It's a region just under your left temple that's associated with processing language, named after the nineteenth-century French anatomist Paul Broca, who discovered it while studying two people with serious speech problems who had suffered damage to this part of their grey matter.

While he's best known for his contributions to neuroscience, Broca also deserves a nod for an important contribution to cancer research: drawing up the first detailed tree of a family affected by multiple cases of breast cancer stretching back three generations. It might seem like a strange side project for someone whose academic interests were mostly focused on everything from the neck up, but it makes more sense when you realize that the family in question was his wife's.

Starting with the grand matriarch of the family, Madame Z, who was born in 1728 and died of breast cancer in 1788, Broca traced sixteen deaths from cancer through the years to 1856. Ten of them were women who died from breast cancer, mostly in their thirties or forties, plus a few who succumbed to cancer of the liver—tumors that Broca suspected might have arrived there from the nearby ovaries. Luckily for Broca's wife, Adele, she avoided her deadly inheritance and lived to the ripe old age of seventy-nine. Less luckily, Broca himself died of a heart attack in 1880, leaving Adele a widow in her mid-forties.

In 1895, nearly thirty years after Broca published his wife's unfortunate family history, a young doctor named Aldred Warthin had a life-changing encounter with a seamstress, Pauline Gross. He had just been made an instructor in pathology at the University of Michigan in Ann Arbor and got chatting to her while taking the long way home through the city's industrious German Quarter. Pauline's grandmother and grandfather had been German immigrants in the 1830s and produced a family with a dismal history of stomach, bowel, and uterine cancers stretching down the generations.

"I'm healthy now, but I fully expect to die an early death," Pauline told him miserably. Warthin's interest was piqued and the two worked together for nearly twenty-five years to pull together her family's unhappy pedigree. By the time

Pauline's grim prediction came true—she died of uterine cancer in 1919, at the age of only forty-six—she'd provided Warthin with detailed medical information about nearly 150 relatives that revealed a clear pattern of cancer inheritance.*

Despite this compelling evidence, Warthin struggled to gain awareness of his work, perhaps because it went against the prevailing ideas of the time that pinned responsibility for cancer prevention and detection on the individual. A message that cancer was hereditary and inevitable was viewed as depressing and doom-laden to the new-founded American Society for the Control of Cancer (now the American Cancer Society). It might have also had something to do with Warthin's enthusiasm for eugenics, which was becoming increasingly distasteful in most parts of polite society.

His work was eventually revisited in the 1970s by American doctor Henry Lynch and social worker Anne Krush, then working at Creighton University School of Medicine in Omaha, Nebraska. Together, they traced more than 650 of Pauline Gross' blood relatives with ninety-five cancers between them, mostly bowel, uterine, and stomach as before. The descendants of "Family G," as they're now known, have since become the longest-studied family in the history of genetics.

Around the same time as Warthin was chasing his way up the branches of Family G's tree, a determined and curious researcher at the University of Chicago was carrying out

* Ami McKay, whose great-great-aunt was Pauline Gross, tells the family's story in her memoir, *Daughter of Family G* (Knopf Canada, 2019).

detailed breeding experiments with thousands upon thousands of mice, showing that some cancers in these animals did indeed seem to be hereditary. The results met with fierce resistance, partly because the prevailing dogma held that cancer was a purely environmental or viral disease, and because there were doubts that the simplistic pedigrees of inbred mice were a good model for the complexities of human inheritance. But some of the pushback also came down to the fact that the researcher in question, Maud Slye, was a woman.

## MICE AND MEN

Slye devoted her life to her mice, even taking them with her across the country by boxcar to California when she had to go and care for her sick mother. She published multiple papers and won several prizes, including a gold medal from the American Medical Association. But she also had to deal with difficult situations such as a completely fabricated story claiming that she'd burst into tears when a male researcher asked to see her lab records, with her vociferous rebuttal of this untruth earning her a reputation for being "indignant."*

Like Warthin, Slye's work on heredity led her toward eugenics, which didn't help with the acceptance of her ideas. She often argued that if cancer is indeed heritable then efforts should be made to breed it out of the human population, pointing out in a talk in the 1930s that "at present we take no

* Slye was a poet as well as a scientist and often reflected the natural world or scientific themes in her work. I particularly love the line "I pace the world because I am storm-driven, By this compelling of creation."

account at all of the laws of heredity in the making of human young. Do not worry about romance. Romance will take care of itself. But knowledge can be applied even to romance."

Even more overlooked than Maud Slye was Massachusetts mouse breeder Abbie Lathrop. Millions of the laboratory mice used all over the world today came from animals originally raised on her farm, and she kept detailed records of her stocks and various breeding experiments. She published a number of scientific papers from 1915 onward about the inheritance of cancer predisposition in mice in collaboration with the eminent pathologist Leo Loeb at the University of Pennsylvania. Yet she never gained acceptance from the academic establishment, who viewed her more as a crazy mouse lady than a serious scientist.

One person who did take notice of all this mouse work was bowel surgeon Percy Lockhart-Mummery at St Mark's Hospital in London, UK, who was busy gathering records from people with familial adenomatous polyposis (FAP)—a rare condition that causes the large intestine to fill with thousands of tiny lumps, known as polyps, almost inevitably leading to death from bowel cancer in the prime of life.

Writing in the journal *The Lancet* in 1925, he described the family trees of three of these patients, drawing the conclusion that both the cancer and the preceding blobby bowels must be inherited. He also noted that:

> Anyone who has taken the trouble to study Miss Maud Slye's work on natural cancer in mice must have been struck by the very remarkable results

> which she has brought forward. These certainly seem to show that there are definite predisposing causes transmitted by heredity which tend to cause mice to die from cancer in succeeding generations.

Despite the bulging drawerfuls of detailed family records and thousands of mice, confusion reigned for decades about the patterns of cancer inheritance and the possible nature of the underlying gene faults. It was very unclear how these family trees fitted together with the parallel world of research going on into carcinogens and viruses in non-inherited, randomly occurring (sporadic) cancers. The increasingly accepted idea of cancer being caused by the gradual accumulation of mutations in oncogenes seemed completely divorced from all this talk of hereditary tumors.

The two strands started to wind together in the late 1960s, thanks to the curious experiments of Oxford University researcher Henry Harris, who was interested in fusing together cells from different animal species to see how they would react. Which genes would they use? What characteristics would these hybrid cells display? And would they manage to divide properly? Harris busied himself blending rabbit cells with rat, humans with frogs, and even the occasional chicken cell thrown in for good measure.

Then the thought struck him. What would happen if he fused healthy cells together with cancerous ones? To find out, he got hold of three different types of lab-grown mouse cancer cells, all of which induced tumors when transplanted into mice, and fused each of them with regular fibroblast cells.

To his surprise, the normal fibroblasts completely suppressed the rogue tendencies of the cancer cells, calming their growth and stopping them from forming tumors when they were put back into the animals.

But this protective ability vanished when certain chromosomes that originally came from the healthy cells were lost (a common occurrence in the uneasy biological truce of a cell hybrid). Harris' conclusion was radical: something in the genes of the healthy cells was acting as a tumor suppressor, normally preventing them from growing out of control. And, more importantly, missing or faulty tumor suppressors would be likely to lead to cancer.

A further clue came from a paper published in 1971 by Alfred Knudson, an American geneticist at the Texas Medical Center who had been studying a rare form of childhood eye cancer called retinoblastoma. He realized that there were two types of children who got the disease. There were those with a strong family history of retinoblastoma, who tended to get lots of tumors in both eyes at an early age, and an unlucky few—just one in every 20,000 children—whose cancer seemed entirely random, only affecting one eye and turning up at a slightly later stage of life.

To Knudson's mind, these childhood tumors didn't fit at all with the prevailing idea that cancer was the result of the gradual accumulation of many mutations throughout life. Instead, he worked out a mathematical calculation to show how both inherited and random retinoblastoma could be explained by faults in just one protective gene, picked up in different ways.

Here's how it works. All our cells contain two copies of each gene—one originally coming from mom and the other from dad. Some of these genes are tumor suppressors, which prevent cells from multiplying out of control. Having one functional copy of a tumor suppressor is enough to keep cells behaving properly, but losing both leads to trouble. Even though he didn't know the identity of the particular gene behind the disease, Knudson proposed that children from families with hereditary retinoblastoma will already be carrying one faulty copy of this unknown tumor suppressor gene in all of their cells. So they only need one more "hit" in the other, healthy copy to lose all of its protective power in that cell, significantly increasing the risk that something will go awry. This means that the chance of developing retinoblastoma is extremely high in these hereditary situations, given the level of mutational shenanigans at work even in healthy tissue. By contrast, a child who has inherited two normal, functional versions of the suppressor gene needs to pick up a hit in *both* copies in the *same* cell. This is a much less likely occurrence, explaining why randomly occurring (sporadic) retinoblastoma is extremely rare in children with no family history.

Together with Harris' discovery of tumor suppressors in the genome, Knudson's Two Hit Hypothesis suggested that cancer was the result of having two missing copies of a protective gene in families with strong patterns of inheritance. But all the work on virus oncogenes suggested that just one mutant copy of one of these drivers was enough to send cells rogue. And what about all the cancers in people with no family history and no viral infection? Nobody seemed to be able to figure out a way of incorporating all of these discoveries into one Big Idea.

Ironically, it was a pair of researchers in Weinberg's oncogene-focused lab who managed to connect all the dots together by finding the elusive retinoblastoma gene—the first known tumor suppressor—which they named *RB*. We now know that *RB* makes a protein that acts as a vital "brake" on the cell cycle, preventing cells from dividing until they're really ready to go. More tumor suppressors appeared over the following years. There was *APC*, another controller of cell growth, which is inherited in families with multiple bowel polyps. *BRCA1* and *BRCA2* were pegged as the culprits behind inherited breast, ovarian, and prostate cancers, and have been shown to have vital roles in repairing damaged DNA. Members of Family G—the unlucky relatives of seamstress-turned-scientific assistant Pauline Gross—carry mistakes in a gene that makes other parts of the DNA repair toolkit. And, of course, there's *TP53*, that infamous Guardian of the Genome. People who inherit a faulty version of the gene have a greatly increased risk of many types of cancer at a relatively young age, known as Li-Fraumeni syndrome.

The final piece of the puzzle came in 1987, when researchers discovered that bowel cancer samples from people with no family history appeared to have a mutation in the exact location in the genome where *APC* was suspected to be hiding. Just as Knudson predicted for children with sporadic retinoblastoma back in 1971, an unlucky double hit in both copies of a tumor suppressor gene could lie behind randomly occurring cancers in people without a genetic legacy of the disease. And the more people started looking in non-hereditary cancers, the more they started to find double faults in tumor suppressors, along with single faults in oncogenes.

Eventually, a picture emerged of two kinds of genes involved in cancer: tumor suppressors, which normally protect and repair our cells, and oncogenes, which normally drive cells to proliferate. It takes a combination of an overactive accelerator and faulty brakes to make a cell run out of control and grow into a tumor.

By the early 1990s, it was possible to draw a mutational map detailing the specific alterations paving the road from normal cell to cancer. The best-defined example was the so-called Vogelgram—first proposed by geneticist Bert Vogelstein—which mapped out five particular mutations that would convert a healthy bowel cell into a small polyp, then a larger growth, a nascent tumor, and eventually an invasive, metastatic cancer.

Work from Weinberg's team supported this idea, showing that they could turn normal human cells into cancerous ones with a combination of five oncogenes and tumor suppressors. Curiously, it only takes two oncogenes to transform a mouse cell, highlighting the difference in cancer susceptibility between short-lived rodents and our own species. As Peto's Paradox would predict, it takes a lot more "hits" to start a tumor in a long-lived, large species like ourselves, compared with a tiny mouse. Humans are remarkably resistant to cancer, although it might not feel that way.

More than a hundred years after the first inklings that

cancer is a disease caused by abnormal chromosomes, we've finally built a picture of how a stepwise combination of faulty oncogenes and tumor suppressors enables a rogue cell to grow out of control and cheat death. Some of these errors may be inherited, providing an unwelcome head start on the process, but most accumulate through life—a collection of typos in the genetic recipe book within our cells, picked up as we age. Damaged cells start to grow unchecked, gathering speed and ticking off a shopping list of mutations along the way, eventually ending up with an unstoppable metastatic cancer.

Back in Boston, Bob Weinberg drains the last of his hot chocolate and ponders a career spent untangling the genetic complexities of cancer:

"I never said we'd find all the genes and I never said we were going to cure cancer—I never imagined that discovering the *Ras* oncogene on its own was going to solve the problem of cancer, I knew it was going to be much more complicated," he reflects in his precise manner. "Still, by 1999 one began to imagine that if we had a repertoire of mutations that would suffice when introduced together to make a normal human cell convert into a cancer cell, that was conceptually very satisfying and it led one to the illusion that maybe we had really solved the cancer problem."

It's an illusion that is rapidly vanishing.

## PATCHWORK OF MUTATION

Non-melanoma skin cancer (NMSC) is a bit of an outlier in the field of oncology. Because these tumors are rarely fatal and easily treated, they aren't even counted in the regular cancer statistics in many countries. Yet they're the most common cancers in the world by far, affecting more than a million people every single year, particularly those with pale skin living in sunny places. This ubiquity and accessible location makes NMSC a useful model for understanding the genetic path that leads to cancer, following the well-trodden idea of a normal cell picking up a stepwise selection of errors in critical genes.

As DNA sequencing technology has evolved to become ever cheaper and faster over the past few years, we now know that cancer genomes are shot through with tens of thousands of mutations and rearrangements. Obviously, not every mutation is important: a gene has to pick up the right kind of fault, whether that's an activating alteration in the case of an oncogene or inactivating a tumor suppressor. And there are a lot of genes and control regions in the human genome that have nothing to do with the kinds of pathways and processes that are involved in cancer, as well as a whole bunch of irrelevant junk DNA where mutations probably don't matter much at all.

Labs around the world have been busy sifting the genetic gold from the junk, creating huge catalogs detailing the mutations in hundreds of thousands of tumor samples. This has revealed hundreds of "drivers"—relevant faults in crucial genes that turn up time and again—with all the other thousands of mutations classed merely as "passengers." The latest research suggests that each individual cancer is driven by up to ten specific faulty driver genes, with certain types of tumor needing more driver mutations to get them going than others. According to our current model of tumor development, these drivers must therefore be the mistakes that are accumulated during the transformation from naive healthy cell to deadly tumor.

Many cancer researchers have gotten very excited about this gigantic molecular stamp collection—particularly because identifying faulty genes and molecules that drive cancer means that pharmaceutical companies can start developing drugs to target them. But looking at the list of all the mutations that have accumulated in a person's tumor only tells us about the end point of the journey and nothing about how it got *here* from *there*. The somatic mutation theory says that normal cells must gradually pick up mutations on the road to cancer, at some point hitting a critical tipping point that inexorably leads to malignancy. Mapping the steps along the way, however, has turned out to be a tricky task.

Some previous studies had shown that samples of normal cells collected right next to a tumor can contain mutations, but this could easily be put down to contamination by escapee cancer cells. Adding to the problem, these observations came from relatively large samples of mashed-up tissue, masking any tiny pockets of cells that might be midway on the path from normal to weird to outright cancer.

Still, looking for mutations in normal cells seemed like a bit of a pointless exercise to many researchers. A century of the somatic mutation theory had embedded the idea that tumors were monstrous mutated versions of normality, so why would anyone look for mutations in healthy tissue?

And it wasn't until recently that DNA sequencing techniques became sensitive and cheap enough to capture rare genetic alterations in tiny samples of normal tissue.

In search of enlightenment, Professor Phil Jones and his colleagues at the Wellcome Sanger Institute in Cambridge set about trying to get hold of some spare skin. More specifically, they wanted to find samples of normal middle-aged skin that hadn't been slathered in sunscreen or hidden under clothes, meaning that the cells were more likely to have picked up detectable mutations from exposure to damaging UV rays at some point.

This seemed like a tricky order to fulfil until they came up with the bright idea of contacting the plastic surgeons at the local hospital, who were routinely removing droopy excess folds of skin from people's eyelids. Whether done for cosmetic or medical reasons, this procedure creates neat little wing-shaped swatches of skin that would otherwise be dumped in the biohazardous waste bin. Importantly, it's a part of the body that gets plenty of UV exposure but is probably the last place anyone would think of smearing sunscreen.

Thanks to the generosity of four people who were prepared to donate their unwanted eyelids to science, the Sanger team was carefully able to dissect 234 tiny samples from these surgical offcuts. Using a highly sensitive DNA sequencing technique, they set about searching for mutations in a hit list of usual suspects—seventy-four of the most common "cancer genes" that had previously been linked with non-melanoma skin cancer and other types of tumor. The eyelid donors ranged in age from fifty-five to seventy-three and all were healthy with no obvious signs of skin cancer, which makes what Jones and his colleagues found all the more surprising.

They'd expected to find at least some the fingerprints of damage from UV light and other sources. After all, this skin had spent more than half a lifetime exposed to the elements, along with all the other general insults of human existence. Instead, they found that the cells were shot through with thousands of mutations—in some cases, as many mutations as have been found in some full-blown tumors. A second surprise was the discovery that these swatches of perfectly normal-looking skin were actually a patchwork made from clusters of mutated cells, known as clones, that had each sprung up from one damaged progenitor. Just a single square centimeter of skin contained around forty different clones, with the largest being around 3,000 cells.

Even more incredibly, they discovered that around a quarter of these clones were carrying faults in cancer driver genes, with some having picked up two or three of these supposedly bad mutations. A number of them had faults in the gene encoding p53, *TP53*, which often turn up in human cancers. Yet, while the somatic mutation "shopping list" theory of cancer might suggest that these cells should be well on the way toward becoming a tumor, they didn't have any outward signs of their inner genetic turmoil. As an example, they found more mutations in a known cancer driver called *NOTCH* in clones covering just five square centimeters of skin than had previously been identified in more than 5,000 cancers in The Cancer Genome Atlas,* the world's largest database of tumor mutations. Would any of these mutated clones eventually have turned into a cancer? We have no way of knowing, but the findings certainly suggest that there are many more "dangerous" mutations in normal tissue than we ever imagined—a chaotic genetic landscape from which cancer eventually emerges.

## WHAT IS NORMAL, ANYWAY?

The discovery that apparently normal healthy cells are a mess of mutations shouldn't have come as such a shock. As far back as 1981, researchers had noticed patches of mutated, potentially cancerous cells in the livers of mice treated with carcinogenic chemicals, which look completely normal under the microscope. Two different research groups have discovered that around one in ten apparently healthy people over the age of sixty-five have cells in their blood carrying the kind of mutations usually seen in leukemia, and another team has found that many cells in perfectly healthy lungs carry faults in one of the most common lung cancer driver genes. Curiously, one study also found a significant number of alleged cancer driver mutations in endometriosis—a non-cancerous but highly unpleasant and painful condition, where rogue patches of uterine tissue start growing in odd places within a woman's body. These wandering uterine cells contain cancerous mutations, which then proliferate

* Abbreviated to TCGA in honor of the four "letters" that make up DNA.

and invade into neighboring tissues yet don't ever become malignant tumors.

An impressive demonstration of the mutational patchwork within healthy tissue came in 2018 from a study led by Phil Jones' colleague at the Sanger Institute, Inigo Martincorena. Moving from external skin to internal tissue, Martincorena had his sights set on understanding the origins of cancer in the esophagus—the tube connecting mouth to stomach. Samples of normal esophageal tissue don't come as handy offcuts from routine surgery, so this time the source was people who had died in accidents or other unexpected circumstances, whose organs were being donated for transplantation to help others.

After obtaining further consent from the donors' families, the team was able to collect small sections of the esophagus from nine healthy people aged between twenty and seventy-five. These samples were then divided into nearly 850 tiny pieces and put through the same careful analysis of cancer drivers that had been applied to the eyelid samples.

Unlike skin cells, which are bombarded by carcinogenic UV radiation, the lining of the esophagus gets a lower exposure to potential carcinogens. And while more than a million people develop non-melanoma skin cancer every year worldwide, the incidence of the corresponding type of esophageal cancer, squamous cell carcinoma, is around fifty times lower. So we might expect that a normal esophagus should have far fewer mutations than normal skin, right?

Wrong. Similar to the eyelids, normal esophagus is also a patchwork of mutated cell clones, but there are many more and they're much larger. Every single one is a cluster of cells that has picked up a different mutation in a crucial cancer gene and started to grow, edging out its normal neighbors. Unexpectedly, although there were fewer mutations overall in the esophagus compared with skin samples from people of the same age, Martincorena and his team discovered a higher proportion of clones carrying faults in supposed "cancer driver" mutations in normal esophagus tissue. Astoundingly, by the time you get to middle age, around half of all the cells in your esophagus probably carry a "cancerous" mutation in *NOTCH* and many others will have faults in other key genes.

To illustrate this observation, their paper contains one of my favorite pieces of genetic data-visualization: the mutated clones of cells are represented as circles of varying colors and sizes, which look like a fabulous retro fabric design. Samples from young folk in their twenties and thirties are scattered with little bubbles of color, shifting to a more closely packed collection of circles for people in middle age, becoming a wild explosion of overlapping ovals fighting for space in a sample from a man in his seventies. Despite almost every single cell carrying one or more cancer driver mutations, he had no outward sign that anything was wrong in his gullet at all.

## WARTS AND ALL

The more scientists look for nasty mutations in normal cells, the more they find. By the time you get to a certain age, every single one of your cells is probably harboring multiple mutations and many of them would be classed as "cancer drivers," were they to be found in a tumor rather than normal tissue. The inevitable conclusion is that all of our cells are a bit messed up. And this is absolutely fine.

"Your skin is a patchwork of mutation," Phil Jones murmurs, peering deeply into my face as we talk. "You have signs of sun damage already, so probably one in twenty cells now has the same mutations you find in skin cancer, just so you know."

"If you live for a really long time you might get two or three non- melanoma skin cancers, and you'll probably get them treated by a nice friendly dermatologist," he adds reassuringly. "But compared with the proportion of those mutations across your skin, that's pretty low. So that's the interesting question—what happens in between a cell having a wonky gene that predisposes it to cancer and ending up with a lump?"

Jones has been trying to answer this question by looking at mice carrying a faulty version of p53 that's often found in cancer, which he can activate on command in just a handful of cells in the skin. This is hooked up to a marker that makes cells glow fluorescent green under a special microscope, so they can spot the progeny of those cells as they start to proliferate. Then they flip the genetic switch to turn on the mutant p53 and patiently watch what happens. To start with, the faulty cells grow fairly quickly, pushing aside their neighbors and turning a small patch of skin fluorescent green. Then the pace becomes slower—after six months there's a noticeably thicker area but after a year it's gone back to normal.

Even more interesting is what happens when Jones takes these animals with their little green patches and pops them under a UV sunlamp, creating the mouse equivalent of a trip to the Costa del Sol. Incredibly, the mutant green cells grow as fast in six weeks as they would in six months, quickly covering areas of several square millimeters. That may not sound like much, but it's a sizeable chunk for a clone of cells on the back of a mouse. While UV light doesn't seem to have an impact on the growth of normal cells, it dramatically accelerates the proliferation of the mutants and creates much larger patches of mutated cells, increasing the chances of one of them picking up a second "hit" in a driver gene and taking another step on the road to cancer. So, you might expect that the longer the mice are left to their fake baking, the more likely these expanding patches would be to develop into little green tumors. But you'd be wrong.

After three months under the UV lamp their skin is peppered with green cells, but by nine months the patches start to shrink and vanish. By using the same DNA sequencing technique that he and his colleagues employed on the human eyelids, Jones saw that the green p53 mutant patches were gradually being crowded out by other clones of cells with even nastier mutations, picked up as a result of the UV exposure, which were well on the way to becoming cancers. The Green Gang may do well when it's the only mob in town, but it quickly gets outcompeted when some new tougher neighbors move in. But this doesn't always have to be a dog-eat-dog situation, as Jones discovered while looking at the patterns of mutations in those original eyelid samples.

"We found a single cell that had colonized over a square centimeter of eyelid, but it had a mutation that activated a gene called *FGFR3*. Why is that so special? Because *FGFR3* isn't a driver of skin cancer and that centimeter of skin will never develop into a tumor."

So, does that mean there might be some mutations that are "good for us," creating clones of well-behaved cells that will never cause any trouble? As Jones explains, activating faults in *FGFR3* are found in seborrheic keratoses—patches of darker, scaly skin that look a bit like waxy warts. They're incredibly common, particularly after the age of fifty, but they're harmless. And once a cell has embarked down the road of becoming a keratosis, it seems to have selected a "safe" evolutionary route that prevents it from going off track and becoming a much nastier cancer. So, I ask him, why can't we use this knowledge to try to find some kind of ointment that might protect against skin cancer by steering cells down this less dangerous path?

"The problem is that your fantasy fancy face cream would just turn you into a massive walking wart," he points out. "But if we could come up with a wonder cream that didn't do that, then it would be a no-brainer."

Jones' work mainly focuses on non-melanoma skin cancer, a tumor that rarely kills. But given that the rest of the body, from which fatal cancers do emerge, is also a patchwork of mutation, then maybe this approach might work there, too. Thinking about ways to support healthy cells to suppress mutant clones—or even just encourage the growth of better-behaved ones—flips the whole paradigm of developing cancer drugs that only target rogue cells completely on its head.

However, we can't rely on the principle that "my enemy's enemy is my friend," and we should be wary of the unintended consequences of attempting this kind of biological engineering. We need to know a lot more about the nature and behavior of the various gangs and tribes in our human landscape before we start interfering with their interactions, lest we accidentally create a deadly monster.

The Sanger Institute team's observation that certain mutations are actually more common in normal tissue than in tumors is also intriguing, telling us that some alterations in so-called cancer genes might actually be protective. For example, an activated oncogene could lead a group of cells to start proliferating wildly but burn themselves out before they can go any further down the road to cancer, by making them divide before they're fully ready.

Another type of mutation might put cells at a competitive disadvantage, making a clone grow less quickly than its neighbors so it gets snuffed out. But the timing is crucial: these kinds of changes would be an advantage if they're picked up later on in the process of tumor growth, enabling cancer cells to keep on plugging away and proliferating as their chromosomes get more messed up and the environment around a growing tumor becomes more toxic.

It's also worth pointing out that the current obsession with targeting the products of driver mutations in tumors might be counterproductive if much of our normal tissue is bristling with these very same faults. Given that Phil

Jones and his team have found that many cells in seemingly healthy skin and esophageal tissue contain supposed driver mutations in the *NOTCH* gene, as well as a significant proportion of tumors, attempting to kill off every one of these mutant cells in a cancer could have serious unanticipated side effects elsewhere.

## WHEN IS CANCER?

The gathering wealth of data showing that healthy and non-cancerous tissues are a mess of supposedly dangerous mutations raises the intriguing question of exactly *what* is cancer? And *when* is cancer? Rather than viewing the body as a static, orderly arrangement of normal cells, a picture is starting to emerge of healthy tissue as a dynamic shifting patchwork of clones in various stages of mutational disarray, expanding and contracting as they shuffle for space and compete against each other. Some mutations might make a clone grow more slowly or even die off altogether while others accelerate expansion, but ultimately the status quo is maintained.

Trouble only starts when a cell picks up a driver mutation and starts proliferating faster than its neighbors, creating a patch of hundreds or even thousands of daughters that all share this same genetic fault. A second driver "hit" is much more likely to happen to any of the cells in this larger expanded area simply because there are more of them than their undamaged neighbors and the cycle repeats again. Hit, expand, hit, expand, repeat until one of those cells has collected enough mutations to make a cancer. This fits with the classic Vogelgram, in which cells must gather a particular set of mutations in order to make the leap toward cancer. And given the trillions of cells in the body, undergoing millions of cell divisions over a lifetime, the emergence of a cancer becomes a numbers game.

Our current idea, built up for the best part of a century, says that cancer starts from a normal cell picking up a simple stepwise accumulation of faults in crucial genes, eventually resulting in a rampant, invasive tumor. But if *all* of our cells are broken and damaged, especially once we reach middle age, then why don't we all get loads of tumors all over the place? What turns a sad cell into a bad cell?

# 5

# WHEN GOOD CELLS GO BAD

Let's do some math.

We're accustomed to thinking that cancer is common. But on the scale of each individual person, it's vanishingly rare. A typical human is made of more than 30 trillion cells, multiplying billions of times over a lifetime. Any one of those could be the one that starts a cancer, but virtually none of them do. According to the somatic mutation theory of cancer, where a healthy cell turns cancerous once it picks up a handful of driver mutations, this would be like a lottery, where your chances of winning are one in ten with fourteen zeroes after it—the same number of stars in a thousand galaxies the size of the Milky Way. We buy millions upon millions of tickets as mutations tick up in our cells over the years, but maybe just one of them will hit the jackpot.

There are a few things that stack the odds in favor of this happening. One is exposing our cells to carcinogenic chemicals or radiation. Another is demonstrated by Phil Jones and

Inigo Martincorena's work, showing that our healthy tissue turns into a patchwork of mutated cell clones as we age. One "hit" in a driver gene might expand the number of cells carrying that fault into a patch ten times the size that it would normally be, increasing the chances that one of those cells might then pick up a second hit that empowers it to outcompete its neighbors and expand into an even more aggressive clone. Hit. Expand. Hit. Expand. Instead of one single cell having to collect a full set of driver mutations, making the odds impossibly small, the task gets easier as each successively mutated patch of cells expands a bit further and faster than its neighbors. It's a bit like being given ten times more lottery tickets every time you match one number, vastly increasing your chances of getting the winning combination.

It all starts to add up pretty fast. A few sunny trips to the beach induce ten mutations per skin cell. Smoking fifteen cigarettes is enough to induce a potentially harmful mutation. Overall, one in five of your cells is probably harboring a mutation in a so-called cancer gene. A smaller proportion have two or even three mutations. Somewhere lurking in your body there are bound to be many tiny tumors that have picked up a full house.

But how worried should we really be? Every one of us probably has a few strange lumps and bumps by the time we get to a certain age, as shown by autopsies carried out on people who have died in accidents. At least a third of women in their forties have a tiny tumor in their breast. Yet only one in a hundred will be diagnosed with cancer at that age and many of them will live their whole lives without a formal cancer diagnosis. The same is true of prostate cancer—something that was first noticed as far back as the 1930s—and far more men will die with the disease than from it. Virtually every single person between the ages of fifty and seventy has a tiny cancer in their thyroid gland, yet just one in a thousand will be diagnosed with a thyroid tumor. Overall, slightly less than half of us are likely to be diagnosed with cancer at some point in our entire lifetime.

There are some other discrepancies in the biological accounting. Small intestine cancers are rare, but tumors in the large bowel are thirty times more common, even though there are very few biological differences between these neighboring regions of our gastrointestinal tubing. If cancer is purely the result of a certain number of mutations in crucial driver genes, then why do liver tumors typically have around four hits while it takes ten to make a uterine or bowel cancer and only one to trigger testicular or thyroid cancer? And if different types of cancer are driven by different numbers of mutations, you might expect some of them to appear earlier in life than others. So why are the chances of developing almost all types of cancer in adulthood fairly low until the age of about sixty, even though our normal tissue is a patchwork of mutation by middle age?

This isn't because we pick up catastrophic genetic errors at a faster rate as we get closer to collecting our pension. It may seem counterintuitive, but the greatest rate of mutations actually occurs in the first flush of youth. Every time a cell copies its DNA and divides there's the capacity for error, which is particularly dangerous if it happens in the stem cells that are responsible for maintaining the body throughout life. The abundant proliferation required to grow from egg to adulthood is absolutely staggering, compared to the daily grind of maintenance throughout the rest of our lives. We grow from a single cell to bouncing, burbling billions in just nine months, adding more and more with every inch ticked off the door frame to adulthood. In fact, you're going to have half the mutations of a seventy-year-old by your eighteenth birthday, but only one percent of the cancer risk.

It's not all about the number of mutations, either. Smokers are much more likely to develop lung cancer than non-smokers, as inhaling all those DNA-damaging chemicals greatly increases the chances of inducing mutations in crucial cancer drivers. So you might expect that smokers would get lung cancer significantly earlier in life than people with the disease who never took up the habit. But you'd be wrong: both groups tend to be diagnosed at similar times of life, mostly after the age of sixty. Smoking strongly influences *whether* or not you get lung cancer, not *when*.

Something doesn't add up.

Tucked away in the Department of Biochemistry and Molecular Genetics at the University of Colorado campus in Aurora, just on the edge of the Rockies, Professor James DeGregori has been working on a theory that explains these discrepancies, which he calls "adaptive oncogenesis." As he sees it, life (at least in terms of cancer risk) is not a linear progression from cradle to grave. Looking at the statistics, the chances of dying from illness at any age from eighteen to thirty-six are pretty much the same. Then things start going downhill and they only get faster as the decades pass. Many researchers focus on the later part of the journey—why we're more likely to get cancer as we age—but as far as DeGregori is concerned, the more interesting question is why we're so much *less* likely to develop it when we're young.

Over the many thousands of years of human history, evolution has shaped our biology to keep us alive as long as we need, but no longer. Virility always trumps longevity: in evolutionary terms what really matters is reproducing successfully. Our species has adapted over millennia to maintain health and suppress the development of cancer throughout the child-raising years of youth and early middle age. Ancient humans with genetic variations that made them more likely to succumb to cancer before rearing children would have been unlikely to pass those unhelpful genes on to the next generation, creating a strong drive toward a population that at least makes it through the reproductive age in one piece.

Coming from the other direction, the dramatic increase in cancer incidence after sixty suggests that however much we love our grandparents and appreciate them helping out with the kids, natural selection has placed a sell-by date on their biology. DeGregori's theory describes how this time limit plays out down at the cellular level. Drawing on more than a century of evolutionary theory, he's switching focus from the large-scale adaptations of species as they've spread around the planet to the evolution of cancer cells within the world of the body.

There are a few key concepts to understand before we get into the weeds of adaptive oncogenesis, which all relate to the idea of evolution by natural selection. The first is this: you have to reproduce, otherwise you can't pass on your genes. Secondly, any population of organisms, from cells to gazelles, cats to bats, and trees to bees will contain individuals with variations in their genetic makeup. There will be plenty of genetic variations that don't really make much of a difference to the chances of reproduction or survival either way and just ebb and flow through the population—something known as genetic drift or neutral evolution.

Very occasionally, a change will be beneficial. For example, a cheetah with a genetic variation that enables it to run a tiny bit faster than its friends is more likely to make the kill and get the main share of the spoils. More food means better health and vitality, making it more likely to get a mate and pass on its quick genes. Over time, more and more animals in the population are likely to have these speedy variants—an example of positive or purifying selection. By contrast, genetic variations that hinder hunting or any other activity contributing to sexual success tend to be weeded out, as they're less likely to be passed down the generations (negative selection).

Finally, natural selection isn't even really about genes or mutations. It's about how well those genetic variations enable an organism to survive and reproduce in its environment—a concept known as fitness. New genetic variants arise all the time, whether through the biochemical rough and tumble within cells or as a result of external DNA-damaging agents like chemicals or radiation. Most will be neutral or negative. Very occasionally, they will be positive. But what counts as positive depends on the environment. Natural selection is sometimes mischaracterized as "survival of the fittest," implying that organisms that are bigger, stronger, or somehow better than their peers will be the ones that make it, but what it really means is "survival of the best fitting."

Importantly, as long as the environment stays the same, there's no selective pressure in favor of any drastic physical changes. Think of so-called living fossils like today's horseshoe crabs, which appear to be virtually identical to their ancestors that trawled the seas 450 million years ago. There's certainly been some tinkering under the hood, genetically speaking, but they're so well adapted to their way of life that any major changes are likely to be detrimental to their fitness and aren't likely to propagate through the population. Natural selection has honed the horseshoe crab to the point where it would take an amazing mutational innovation to improve on half a billion years of Just Right. In evolution-speak, this would be described as reaching a "fitness peak." Or to put it more simply, "if it ain't broke, don't fix it."

The definition of fitness for any given organism depends entirely on its environment. Blind cave fish don't last very long in brightly lit open waters, but they thrive in the dark holes that sighted predators tend to avoid. Fur-coated polar bears are perfectly adapted to a life of hunting and breeding in the frozen Arctic but are struggling to survive as the climate warms around them. Evolution is most noticeable when the environment changes and fast shifts in conditions make for rapid selection (or rapid extinction).

According to the adaptive oncogenesis hypothesis, we can understand the emergence of cancer using exactly the same principles. Over hundreds of thousands of years, stem cells in the organs of the human body have become supremely well adapted to their particular biological landscape, from the smooth flatlands of the skin or fast-flowing soup of the bloodstream to the spongy lung and undulating gut. Crucially, evolution has shaped these cells for maximum fitness in the spritely, pert environment of a young body. Like horseshoe crabs, physically unchanged for millions of years because they're so well suited to their niche on the sandy shore, there's strong pressure for stem cells to maintain the status quo in our tissues as long as this youthful environment stays exactly the same.

As we go through life, cells pick up their own unique collections of mutations, creating a patchwork of genetic diversity that acts as fuel for natural selection. Because stem cells are so well adjusted to the young tissue environment, most mutations are likely to have a neutral or negative impact on fitness, causing damaged cells to proliferate more slowly than their healthy neighbors or just to give up and die. Very occasionally, there might be a beneficial mutation that gives a cell a slight edge over its neighbors, enabling it to multiply a bit faster and start expanding its patch—the kind of change that would signal a step along the road to cancer—but these mutant cells are going to have to be supremely fit to compete with young stem cells on their home turf. We know that this must occasionally happen, as people are occasionally diagnosed with cancer in their twenties, thirties, forties, and fifties, but these cases are often linked to an inherited gene fault, which puts every cell in the body one step further along the journey to cancer.

As we get older, things start to change. Entering my fifth decade, I'm acutely aware that things are starting to fall apart. I'm not just talking about gray hairs, sagging boobs, and increasing anxiety about whether or not I can still get away with wearing a short skirt—I'm talking about what's going on at the cellular level as we age. We live in our bodies for eight decades or more, if we're lucky, and things will inevitably start to look a bit shabby.

As life goes on, our tissues and organs accumulate damage. Despite the best efforts of our cellular repair kits, mutations build up that change the molecular makeup and behavior of cells and the habitats in which they live. These alterations don't just increase the opportunities for potentially cancer-causing faults in key driver genes. They also affect all the other genes that are important for regular cell function and repair too, disrupting the levels of cellular signals, interfering with the molecular Velcro that holds cells together, disrupting the levels of hormones and generally making a mess of the place.

The usual processes of cell maintenance gradually wind down with age, especially after we've passed our prime reproductive years. For example, the cells in young skin are tightly connected together, preventing any would-be tumors from staging a takeover and eventually resulting in the elimination of rogue cells. But the connections within our tissues become looser over time, making it easier for rogue cells to start moving around as a cancer grows and spreads. And as well as causing damage to DNA, carcinogens like tobacco smoke and UV light damage the stretchy, supple collagen molecules that bind our cells together, giving even more opportunities for cells to stray.

The gradual degradation of aging even extends to how our genes are packaged up and used. Young cells have neatly packed DNA, coiled tightly around ball-shaped proteins called histones, which are labeled with various molecular tags that are involved in controlling gene activity, known as epigenetic modifications. This orderly arrangement starts to go awry in aging cells. The DNA coils start to unwind and the modifications are disrupted, increasing the chances that genes will be switched on or off when and where they're not wanted. Even down at the level of the genome, everything gets a bit saggy as we get older.

One of the most significant manifestations of aging isn't gray hair, wrinkles, or baggy chromosomes but inflammation. In 1863, German pathologist Rudolf Virchow noticed the presence of leukocytes—a type of immune cell—nestling in between the cancerous cells in a tumor, leading him to propose that cancer arose as a result of "constant irritation," causing activation of the body's immune defenses.*

* As well as being one of the early pioneers of cancer research, Virchow also worked on roundworm infection in undercooked meat, which was widespread in Germany at the time. He was a political liberal and his views on decent living conditions, education, and health for the poor got him on the wrong side of the conservative Prussian chancellor, Otto von Bismarck, who was so irritated he challenged the scientist to a duel. According to legend, Virchow's weapons of choice were two identical sausages, one of which was liberally seasoned with roundworm parasites. Fearing the Wurst, Bismarck refused the duel. (While this is a great story with an excellent joke, it's sadly not true. Although Bismarck did challenge Virchow to a duel, the scientist just turned him down.)

His notion was derided as nonsense and ignored in favor of the line of thinking that would eventually grow into the somatic mutation theory (page 66). More than 150 years later, researchers are wising up to the idea that Virchow may have been on to something, after all.

Inflammation is the most obvious manifestation of the immune system at work. Virtually everyone will have experienced acute inflammation at some point: the throbbing hot redness, swelling, and unpleasant seepage marking out the site of an infection or wound. Acute inflammation is a life-saving explosion of biological activity, calling in the immune cell cavalry to destroy invading microbes or infected cells, and increasing the flow of blood and nutrients to help with repairs. But we may be less aware of chronic inflammation—a slow, smoldering immune response persisting over months or years which is associated with diseases ranging from heart disease or diabetes to—yes, you guessed it—cancer.

Chronic inflammation can be caused by persistent infections, long-term exposure to irritating chemicals, and autoimmune diseases where the immune system starts to attack healthy tissue. There are also more mundane triggers, the biggest unavoidable one being age. As we get older, the levels of chronic inflammation in our tissues creep up. This is the inevitable result of collateral damage from the biochemical processes at work inside our cells, the gradual accumulation of harmful chemicals in the body, a lifetime of infections and afflictions, and the general downhill slide as time ticks by. It may also be related to declining levels of sex hormones, such as estrogen and testosterone, which help to keep a lid on inflammation in our younger years. As might be expected, smoking is a potent cause of inflammatory damage in the lungs and also damps down the body's anti-inflammatory responses. Keeping hold of excessive body fat is another risk factor. Rather than being inert flab, fat storage cells produce many molecules that fuel chronic inflammation.

Another potential but underexplored cause of chronic inflammation is stress. The idea that stress causes cancer looms large in the popular imagination, but most studies don't show any link between specific stressful life events such as bereavement or divorce to an increased chance of developing cancer. However, there may be a connection with long-term chronic stresses such as financial insecurity and poor housing, which are more likely to be experienced by people who are less well-off. Health inequality is a well-known issue, with people at the lower end of the socio-economic ladder more likely to become sick and die younger from all kinds of ailments, including cancer. This discrepancy tends to be blamed on the usual suspects—bad diets, obesity, smoking, and drinking—but these factors don't add up to the full story.

In a study of more than 8,000 people across the whole social spectrum, researchers at the University of Essex in the UK have carefully measured the levels of two molecules in the bloodstream that are associated with long-term inflammation and can be exacerbated by stress. They found that the levels of these chemicals rise more sharply from the age of thirty in people in lower income groups, peaking in middle age. By old age, the gap between rich and poor narrows again, suggesting that the inflammatory effects of aging catch up with everyone in the end, no matter how wealthy they are.

Of course, there's no one social group that has a monopoly on stress, but the Essex team's results reveal an increased level of exposure to chronic inflammation throughout the prime of life in people who are stressed out by debt, insecure and dirty housing, and everything else that comes with financial insecurity. Lack of sleep adds another strand to this Gordian knot: poor sleep is linked to increased levels of chronic inflammation, and goes hand in hand with anxiety, stress, and difficult living conditions. Cancer risk is a social issue as well as a biological one and it's an area that urgently needs more research.

Whatever the cause (and there are likely to be many contributing factors in every person), chronic inflammation creates a disturbance in the cellular habitats within our tissues, producing an environment in which mutated cells are more likely to thrive. By way of demonstration, DeGregori and his team have done some clever experiments with mice that have been genetically engineered to carry the same driver mutation that's seen in non-smokers with lung cancer.

Flick on the mutation in young or middle-aged mice and nothing much happens. Plenty of lung cells have the faulty gene but only a handful of them even get as far as creating little precancerous lumps called adenomas. Do the same thing in old mice and their lungs start to fill up with adenomas, any of which could be on the path to becoming a cancer. Adding in a second gene encoding a potent anti-inflammatory protein makes a big difference, bringing the number of adenomas in the elderly mice back in line with the young 'uns. Apparently, these animals also look younger and sleeker in their dotage too, highlighting the connection between inflammation and aging.

There's plenty more evidence assembling for the role of chronic inflammation in altering the landscape of the body and encouraging the growth of cancer. Some inflammatory conditions increase the risk of developing cancer, such as hepatitis and Crohn's disease, although others don't—inflammatory asthma doesn't seem to have an impact on lung cancer risk, for example.

Large-scale trials of the oldest anti-inflammatory drug, aspirin, have shown that long-term use over a decade or more can reduce the risk of bowel tumors and other types of cancer. Unfortunately, despite being cheap and widely available, daily doses of aspirin come with potentially deadly side effects, including stomach bleeding and strokes, so it's not something to try without medical advice.*

Of course, we can't get rid of inflammation altogether. Acute inflammation serves a vital healing purpose, and it would be a spectacular own goal to block all inflammatory pathways in the hope of preventing cancer, only to die of infection from a grazed knee. But figuring out safe ways of preventing or damping down the smoldering burn of chronic inflammation without hamstringing the important bits of the immune response could be a very useful way of stabilizing

* One researcher told me that all the oncologists he knew were taking small daily doses of aspirin, while all the gastroenterologists weren't.

the habitats within the body and keeping wannabe cancer cells under control.

It's not just aging and inflammation that shift the selection pressures and promote the growth of cancer. Treatments like surgery, radiotherapy, and chemotherapy cause significant damage to the tissue in and around a tumor, creating a post-apocalyptic world where the normal rules no longer apply. Developing treatments that help to restore the micro-environment after cancer treatment could help to reduce the chances of any weird stray cancer cells finding a suitably messed-up niche in which they can start growing again. Studies in mice have shown that even something as simple as giving anti-inflammatory drugs following breast cancer surgery might make a significant difference to the chances of the cancer coming back—an important result that deserves a lot more investigation.

## PEER PRESSURE

If we return to the idea of the "society of cells," we can imagine the environment inside a young body as a highly ordered and well-policed civilization. The molecular streets are kept tidy and well maintained, every cell knows its job and its place in the world, and any wrongdoers are taken out by the immune system. Only stem cells proliferate, making new cells as and when they're needed, while the rest stand down (quiescence), only springing into action in a state of emergency like an injury. Faulty and damaged cells die (apoptosis), while older cells that are no longer needed will quietly sit on the biological equivalent of the porch in a rocking chair, watching the world go by and minding their own business (senescence).

The stabilizing and restrictive influence of young tissue must be incredibly potent, given that even something as damaging to cells as smoking doesn't seem to bring forward the onset of cancer too drastically. Unfortunately, nothing lasts forever. If young tissue is like a strictly regulated society, where nothing is allowed out of place and dissent is not tolerated, then the inside of an older body starts to look a lot more like a free-and-easy neighborhood where anything goes. And, as is the case for evolution everywhere else in the natural world, once the environment changes, the pressures of natural selection change, too.

The discovery of a surprisingly high number of what should be considered "cancer driver" mutations in young and middle-aged people (page 109) tells us that genetic damage will always happen, from cradle to grave. But while potentially cancerous mutant cells can't compete with fit cells in the orderly buff environment of a youthful body, their strange genetic quirks give them a fitness advantage in more disorderly older tissue. A shipshape society might suit law-abiding youthful cells, but weird, old mutated cells are better suited to life in a weird, old mutated world. What might be considered antisocial behavior in a young environment is tolerated or even encouraged as the rules governing our cellular societies start to break down, making it more likely that cheating cancer cells will emerge and prosper.

The idea that the body's internal environment can have

a powerfully suppressive effect on cancer dates back to at least the 1970s, when embryologists Beatrice Mintz and Karl Illmensee discovered that cancer cells mixed in with the healthy stem cells in a mouse embryo could be persuaded to fall back in line and play a normal part in development, even going on to become eggs and sperm that gave rise to perfectly healthy pups.

A couple of decades later, the pioneering American–Iranian cell biologist Mina Bissell showed that breast tumor cells would knuckle down and play nice if they were placed in the company of well-functioning healthy cells, but reverted back to their cancerous ways once this repressive environment was removed. Flipping it around, good cells will go bad if they fall in with a bad crowd and previously healthy cells will turn rogue if they're placed in tissue that's been treated with a carcinogenic chemical. But these results—and the notion that cancer emerges from the breakdown of cellular society rather than the accumulation of specific mutations—were largely forgotten or ignored in the rush to Sequence All the Genes.

To put it simply, cancer doesn't start when a cell picks up a certain number of mutations and grows out of control. It starts when a cell picks up faults that enable it to ignore the usual rules of multicellularity, becoming fitter than its neighbors and outcompeting them. As the cell biologist Harry Rubin succinctly put it, "Cancer is the result of an inexorable process in which the organism falls behind in its ceaseless effort to maintain order."

Some cells achieve fitness through genetic changes, while others have fitness thrust upon them. A particularly nasty combination of mutations could give cells a strong fitness advantage over their neighbors in either young or old tissue. Alternatively, a group of unhappy cells might suddenly find they're the fittest cells in town if every cell around them is doing even worse—a three-legged horse can't win a race against four-legged steeds, but it'll be a stand-out favorite in a field of two-legged nags. However it happens, rogue cells that are best adapted to the shabby aging environment are more likely to survive and proliferate, going further down the road to becoming a tumor.

As an aside, it's worth remembering that most of the cancer research in animals has been done using young male mice. It should hopefully be clear to you why this is a terrible idea and maybe explains why so many exciting new cancer drugs fail when they get into clinical trials. Not only do mice have a completely different life history strategy from humans, living fast and dying young, spritely youthful animals aren't a remotely good model for the aging tissues of elderly human cancer patients.

This may go some way to explaining why we're so good at curing cancer in mice, and so bad at achieving the same outcomes when we try to transfer our bright ideas into humans. Instead of testing every promising drug in two different animal species, as is currently required (usually a rodent then a non-rodent, such as dog, primate, or pig), perhaps the regulatory agencies should ask to see data from young and old animals instead.

If we view the emergence of cancer in old age as something that's hardwired into the evolutionary history of our species—adaptations that have been selected for getting us through reproduction and not much further—then it also suggests that there's a finite length to human lifespan and we can't evolve our way out of it in a hurry. This is a hotly debated topic in aging research, with some arguing that human life expectancy should top out somewhere around 120 years. Others maintain that there's no fundamental biological limit on lifespan and a whole longevity industry has sprung up arguing that the first child to live to 500 or even 1,000 may have already been born.

Curiously, a study of elderly people in the American state of Utah showed that while the risk of developing cancer increased up to the age of ninety, it actually went down in people lucky enough to live longer than that. There's no clear explanation for this peculiarity, although it may be the result of a gradual slow-down of cell division with age. Perhaps by the time you've made it into your tenth decade, your stem cells are dividing so slowly that they're no longer capable of growing into an aggressive tumor even if they've acquired the mutations that would enable them to do so.

Personally, I think that there's a huge evolutionary boundary to leap before there will be a significant change in the top end of human lifespan. If women continue to have children later in life, then we might start to shift our whole reproductive life history toward longer fertility and longevity. But human evolution is a complex interaction between genetics and environment conducted over tens of thousands of years, so it's difficult to catch it in action. Unless something incredibly drastic happens to the world to exert an enormous selective pressure (hello, climate change!), then I'd be very surprised if we make the leap to living to 500 any time soon.

I very much enjoy reading the musings of certain transhumanists who argue that there's a straightforward multistep plan to achieving immortality. First on the list is the trivial matter of curing cancer, which is usually viewed as a hefty but fixable engineering problem. If I've learned anything about biology over the years, it's that it's a lot more squishy and unpredictable than any engineer could possibly imagine.

Whether we like it or not, humans come with a built-in shelf life. We've evolved to maintain our tissues in good health for a limited time only, after which everything just seems to give up and the bad guys take over, like gangs moving into a run-down city. The obvious solution to this problem would be to keep on repairing and rejuvenating our tissues forever, maintaining the youthful environment that favors healthy cells and discourages cellular misbehavior. This is the biological equivalent of New York City's broken windows policy, laying down the firm hand of the law to restore order and encourage good citizenship. If we could stop the slide into decrepitude or even reverse the gradual decline of our tissues with age, then there's a good chance we could drastically reduce the incidence of cancer (and maybe also look younger in the process).

## THINK OF THE CHILDREN

A note here about cancers in children. There are few things more distressing than childhood cancer. We can just about deal with the idea that we might grow old and eventually die, hopefully after living a long and happy life, but the sight of a toddler's belly swollen by a grotesque kidney tumor or a schoolkid grappling with leukemia alongside their SATs seems inexplicably cruel. It also runs counter to the narrative of cancer emerging as a result of mutated cells evolving their way out of the constraints of aging tissue. This is because childhood cancers are fundamentally different diseases to adult tumors.

I spent countless hours of my PhD staring down a microscope watching life unfold, monitoring early stage mouse embryos cycling through their exponential launch sequence. One. Two. Four. Eight. Sixteen . . . At some point I'd lose count as the tiny ball of cells began to hollow out and reorganize into an empty sphere with a minuscule clump of almost invisible stem cells stuck somewhere on the inside. If transplanted back into a female mouse, this microscopic football would embed into the uterus. The outer cells would start to build a placenta, while the stem cells would continue to multiply and specialize, growing into what would eventually be born as a squeaking pink pup. Even though I watched more embryos than I can possibly remember, I never stopped being amazed at the capacity of this tiny handful of near-magical cells to generate all the tissues of an animal's body, from the whiskers on a curious sniffing nose to the tip of a wriggling tail.

Childhood cancers are the result of these normal developmental processes running awry. Solid cancers like Wilms' tumor (a type of kidney cancer) or neuroblastoma (cancer of the nerve cells) arise from cells getting "stuck" at a certain point in development, continuing to proliferate instead of specializing and settling down where they're supposed to. Each type of tumor turns up in a specific location or cell type at a particular time with a characteristic set of mutations, depending on the particular developmental roadblock it has run up against. Leukemia in children seems to be the result of a "two hit" process, the first being some kind of inherited genetic mutation and the other being exposure to infections in the first few years of life.

However they occur, childhood cancers are mercifully rare and the chances of a cure for many types are now quite high. The flip side of the success is that more and more kids are having to live for decades with the long-term side effects of treatment, which may mean infertility, hearing loss, memory problems, and more. It's a tough job persuading researchers and pharmaceutical companies to get interested in such a small population of patients, compared with a potential market of millions of adults, but a lot more could be done for the cancers where there are no effective treatments and to make the cures we do have a lot kinder.

## BREASTS, BOWELS, AND BAD LUCK

Somewhere in a hospital freezer in East London are the remnants of my friend Desiree's breasts, waiting for someone to convince a graduate student to investigate her family's unusual *BRCA2* mutation. A genetic test had already revealed that she'd inherited one copy of the faulty gene responsible for breast cancer in her mother, grandmother, most of her grandmother's sisters, and many of their daughters. But by the time she got around to thinking about scheduling a preventative double mastectomy it was too late.

Desiree* was part of a close-knit gang I worked with at Cancer Research UK, and sitting alongside her as she went through her diagnosis, treatment, and recovery from breast cancer was a sobering experience. Still, we did buy her one of those plastic clown hats with fake hair as a joke replacement for her vanished locks, because we were terrible people who didn't know how to tell someone we loved that we were really quite worried about them.

When I recently caught up with her, she reminded me of one of the most curious and enduring mysteries about the disease. Every cell in her body is carrying a broken copy of the *BRCA2* gene, which she got from her mom. And of all the trillions of cells within her body, a significant proportion of them will have also accidentally damaged the functional *BRCA2* backup copy that she got from her dad. All those "two hit" cells have seriously impaired DNA repair toolkits. So why did she only get cancer in one breast? She's not unusual, either. The range of cancers that tend to turn up in people with hereditary *BRCA1* or *BRCA2* mutations is remarkably small. Tumors in the breasts and ovaries for women. Prostate cancer for the guys and breast cancer too, although much more rarely. Sometimes pancreatic cancer, in either sex, and maybe brain tumors. But they don't seem to be at increased risk of getting lung, bowel, or any other type of cancer than anyone else.

There are other oddities. Families with inherited faults

* Her name has been changed for privacy.

in a gene called *APC* have a condition where their bowel blooms with thousands of tiny little lumps, any of which might develop into a tumor if left untreated. There's also an increased risk of liver and thyroid cancers, but that's it. Broadening out to the general population, we don't understand why lung cancers tend to have mutations in their *EGFR* or *ALK* genes, while melanomas usually have faults in *BRAF*. Why are breast and prostate tumors so common, while heart cancer is vanishingly rare? We don't really know.

The most promising explanation lies in understanding how all the tissues of the body end up being so different. All your cells contain the same collection of 20,000 genes, but they aren't all switched on all the time. For example, a liver cell has to activate the subset of genes that do liver-specific stuff—making digestive enzymes and so on—and switch off all the irrelevant ones. A brain cell needs to turn on the instructions for making neurotransmitters but shouldn't ever switch to making muscle.

These patterns are the end result of a long series of decisions that are made on the journey from single cell to embryo to baby to adult as cells proliferate, migrate, and specialize, responding to signals and cues from the cells around them. Maybe some of these developmental pathways and choices make it more likely that certain mutations will occur or make it easier for cells to ignore the rules that govern their biological society.

Then there's the matter of how the cells in each tissue are arranged. It's notable that most cancers start in the linings of the ducts and tubes in the body where they have space to expand—such as the ducts in the breast or prostate gland, the branching tubes in the lungs, and the sewage pipe of the gut—rather than more tightly packed tissues like the pancreas or brain, which have lower cancer incidence. Another factor is the pace of cell replacement. Bowel cells only last a few days before they're shed into the gut and pooped out, but that means that there needs to be a big pool of stem cells, any one of which could turn cancerous under the right conditions.

Shedding millions of cells into the toilet every day is a pretty good cancer-prevention strategy—if a rogue cell is no longer in your body, it can't grow into a tumor—but it's not infallible. Bowel cancer is still one of the most common tumors in many parts of the world. By contrast, there are very few stem cells in the heart and they have limited proliferative abilities. This is useful for reducing the risk of heart cancer to nearly zero but garbage for repairing the effects of a heart attack.

The rate of turnover probably also has an impact on how much effort a particular tissue puts in to repairing mistakes in DNA—there's little point in fastidiously correcting every error in a cell that's only destined to live for a few days, but it's more worthwhile if it's got to stick around for years. Fascinatingly, plants appear to do the same thing, accumulating more mutations in short-lived disposable petals compared with longer-lived leaves and stems. And African killifish, which only live for a year, are happy to pile up huge numbers of mutations in their DNA.

Another explanation is sheer bad luck. In 2015, mathematician Cristian Tomasetti and cancer genetics legend Bert Vogelstein at The Johns Hopkins University School of Medicine in Baltimore published a paper that created a bit of a stir, to put it mildly. The pair had been wondering why some bits of the body that are constantly churning out new cells, such as the bowel or skin, are much more likely to spawn a tumor than, say, brain or muscle. To find out, they mapped the proportions of tumors arising in more than thirty different body parts in American cancer patients against information about the number of stem cells in each organ and how often they divide. Their conclusion was that two thirds of the variation in cancer rates between tissues is purely down to "bad luck"—a chance glitch resulting from the basic processes of proliferation inside cells, such as a mistake in DNA copying or repair—while only a third of the variation could be blamed on environment, lifestyle, or inherited gene faults. Simply put, the more that the cells in a particular tissue are multiplying, the more likely it is that something will go wrong and send them down the road to cancer.

It's important to point out that this is about body parts, not people. The paper doesn't say that two-thirds of *cancer cases* are the result of bad luck—a widely repeated misinterpretation of their results. Rather, they suggested that inherited mutations or external factors like toxic chemicals play a relatively small role in determining how cancer-prone any particular organ is, compared with the much greater influence of unavoidable processes of life.

To put it simply, tissues that are packed with stem cells that multiply rapidly are more likely to get cancer than those with a slower turnover, because more cell division means more chances for stuff to go wrong. Adding in the effects of environmental, lifestyle, or inherited factors makes a relatively small difference on top of the risk that's already there owing to the underlying rate of mutation resulting from this cellular churn.

There are a few notable exceptions that stand out from their data: the influence of smoking in lung cancer, HPV in head and neck cancer, and hepatitis C virus in liver cancer, plus two forms of bowel cancer that are strongly associated with inherited gene faults. But there seemed to be little association between melanoma and external causes, which is most curious given that Mike Stratton and his team find the fingerprints of UV damage all over the DNA from these skin cancers (page 76).

Regardless of the scientific subtleties, most of the news coverage ran with headlines such as "Two-thirds of adult cancers largely 'down to bad luck' rather than genes" and "Most cancer is beyond your control." To make matters worse, the story broke on New Year's Day—a time when most people are sweeping the Christmas empties into the recycling bin and steeling themselves for a year of healthy resolutions, and bored journalists are desperate for stories. *See?* went the argument. *Why bother ditching the cigarettes, throwing out the booze, or hitting the gym when it's all down to bad luck?**

* When I spoke with him about the controversy, Tomasetti maintained that he and Vogelstein clearly showed that known risk factors such as smoking caused a big increase in the chances of developing certain cancers, rather than implying that attempts at healthy living are a waste of time. "I don't know about for you, but for me when someone tells me that if I don't want to get cancer in my life there's a factor that accounts for a third of all cases, it would be irresponsible to say I don't care about that," he told me. "Those are still good odds, so you definitely want to

As well as creating a media shitstorm, the "bad luck" story unleashed a tsunami of controversy in the cancer research community, triggering more than a hundred critical responses. Many of the scientists I've interviewed for this book have ventured an opinion on the paper, often unsolicited and almost universally negative. I've heard criticisms of everything from the estimates of cell numbers and choice of data sets to the statistical methods and analysis, with one person who shall remain nameless describing it as "complete and utter bollocks."

In response, Tomasetti and Vogelstein doubled down, publishing a new paper in 2017 using revised estimates to come to more or less the same conclusion. Yet again, their findings provoked another round of misleading headlines and ensuing scientific backlash. One particular area of dispute centers on estimating the number and proliferation of stem cells in using data from mice rather than humans. For a start, mice have a "live fast, die young" strategy compared with the "slow and steady" human approach, so their tissues have evolved to turn over in a very different way from ours. The rates of cancer in various tissues is also wildly different in mice versus humans, with our furry friends being much more likely to develop cancer of the small intestine compared with the large bowel, while we're the other way around.

Then there's the work done by stem cell researcher Ruben van Boxtel and his team at the Princess Máxima Center in the Netherlands. They've discovered that the human liver, small

---

reduce the risk as much as you can."

intestine, and large intestine (bowel) all accumulate mutations at the same rate—roughly forty DNA "typos" per cell per year—yet liver cancer is nine times more common than small intestine cancer, and bowel cancer is twenty-eight times more likely to occur than a tumor in the small intestine next door. So there has to be more to cancer risk in each of these tissues than the simple tick of mutations over time.

Another problem with the Bad Luck paper is the assumption that all organs in the body work the same, with a fixed pool of stem cells responsible for maintaining that tissue. But this simply isn't borne out by biology. We don't know what a stem cell really is anyway, and there's a lot of argument about whether cancer arises from normal stem cells in the body that have picked up mutations and gone bad, or if more specialized cells can revert to a more stem-like state once they accumulate enough genetic hits.*

There are what could be termed "professional" stem cells in the intestines, which keep on proliferating and renew the entire lining of the gut on a weekly basis. Even in the rigid organization of the bowel, where stem cells are constantly churning out specialized lining cells that can no longer proliferate and will eventually be pooped out, cells that have started this fateful journey can turn around and become stem cells again if their original progenitors are sick or dying.

Conversely, nobody has ever successfully managed to find anything that looks like a liver stem cell. In fact, you can have

* The answer seems to be "It depends on the tissue and the tumor," which isn't terribly useful.

two thirds of your liver surgically removed and it will grow back. This isn't the work of pro stem cells but "amateur" liver cells moonlighting as stem cells when they're needed. It's starting to become clear that being a stem cell is a state, switched on or off depending on the context and environment, rather than a fixed fate. So it's hard to see how Team Bad Luck could really be sure of their estimates of stem cell numbers in the first place.

There was also some head-scratching around the selection of cancers, particularly the omission of breast and prostate—two of the most common tumor types. The sole focus on cancer cases in the USA ignores the massive variation in incidence of certain cancers depending on where in the world you live or come from. One notable example is a classic study from 1973, showing that incidence of breast cancer jumps for women who move from Japan to California even within a single generation, and I doubt that they underwent an enormous rearrangement of their fundamental biology when they crossed the Pacific.

Scientific arguments aside, there's a more subtle point here. Decades of public cancer-prevention messaging has been built on telling people what to do—or not do—to stay healthy. Don't smoke. Don't drink too much. Eat your greens and don't be fat. A parallel narrative pins the majority of blame for cancer on malign external forces that must be avoided at all costs, ranging from the sensibly evidence-based (air pollution or toxic industrial chemicals, for example), to the conspiratorially ridiculous, such as wind turbines, chemtrails, the 5G network, and more.

The announcement that a significant chunk of our cancer risk is purely chance represents a significant loss of control over our own lives and health—ironically turning the blame back onto the gods or fate rather than scientifically proven and preventable causes. This might be less bothersome for statisticians who deal in gray probabilities rather than black-and-white certainties every day, but it's profoundly unsettling for the rest of us.

The simplistic competing stories of "Your cancer was caused by X" and "It's just bad luck" are comforting but ultimately unhelpful. It's reassuring to tell ourselves that a deadly tumor turning up in someone who seems to be a perfectly healthy paragon of wellness is bad luck, pure and simple, while lung cancer in a lifelong smoker is hardly a surprise. But this kind of narrative leads to confusion, blame, and guilt.

I will never forget having to answer a letter from a woman whose husband had died of pancreatic cancer who had written to our team at Cancer Research UK after seeing that our website stated that processed meats increase the risk of the disease. She'd sent her husband to work every day with a ham sandwich and was distraught at the thought she might have inadvertently killed him with kindness and charcuterie. So many patients and families want to know the answer to the question "Why me?" yet in most cases the best we can do is offer a list of prime suspects with varying degrees of culpability.

After spending time talking with Christian Tomasetti, I can't help feeling that he's come to the correct conclusion via the wrong route. There's an important scientific and public conversation to be had about what genuinely causes cancer, the things that increase the risk, and what we can (or can't) do to prevent it. Right now, the vast majority of funding goes toward research leading to new treatments—whether that's lab studies, drug development, or clinical trials. A smaller piece of the pie is spent on developing new tests to diagnose tumors at an earlier stage, when they might be more successfully treated, and the paltry leftovers go toward prevention. Given the massive impact that preventing cancer might have on extending healthy lifespan, compared with the current futility of treating advanced disease, I call bullshit.

Effective prevention must come first. In 2018, 17 million people around the world were diagnosed with cancer, with nearly 10 million of them dying from the disease (and those are the ones that we definitely know about). Cutting that by just 10 percent would make a huge impact on society, reducing personal pain and misery and significantly cutting healthcare costs. Eradicating all cancer-causing infectious diseases would make a big difference, although it's a topic that's rarely talked about in places like Europe and North America.

Worldwide, one in every five cases of cancer is linked to viral, bacterial, or parasitic infections, particularly in less wealthy countries. Bringing in widespread treatment or vaccination to control these diseases, or developing them where none currently exist, would make a significant dent in the global cancer statistics and, more importantly, reduce a great deal of human misery.

Another big one is tobacco control. Sorry if it sounds boring, but it's true. Globally, tobacco causes around 7 million deaths every year from cancer and other diseases. That's nearly 20,000 every single day, or the population of a reasonably sized town wiped out in a week. I was lucky enough to speak with Richard Peto at the University of Oxford (he of the eponymous Paradox), where he worked with the late Richard Doll to confirm the links between smoking and disease. Now in his mid-seventies, and remarkably spritely after being treated for advanced intestinal cancer the previous year, I asked him how he felt about the tobacco companies for causing so much human misery over so many decades.

"There's no point getting angry about it, just as there's no point getting angry with bacteria for becoming antibiotic resistant, because what can you do to prevent resistant strains emerging? You need to work out what regimens minimize the risk," he told me. "There's no point being angry with tobacco companies because they can perfectly well say, 'If we didn't sell them somebody else would.' And you could say the same about the marketing people—if they weren't trying the most devious ways they could find for marketing cigarettes then the company should fire them and get another PR company."

Tobacco companies have emerged as powerful merchants of death in a global society where financial returns to shareholders trump human health. Nicotine is a highly addictive, widely available, enjoyable, and cheap drug, and people aren't going to give it up in a hurry. Countries that are serious about reducing cancer rates will have to create an environment where the balance of power and money is shifted away from tobacco companies in favor of health. So what works?

"The rough rule is: trust the tobacco industry—whatever they think is important probably is," Peto says. "If they really, really don't want an advertising ban, then ban advertising. If they're quite happy to support educational campaigns in schools then these probably aren't much use. They really hated plain packaging—which tells you something about that—but the single biggest thing is price."

Showing me a graph of tobacco consumption in the UK, he points out the modest impacts of advertising bans and smoke-free legislation, which restricted smoking in enclosed public spaces. But the two biggest falls in smoking by far have come when the price of cigarettes has gone up. The first was in 1947, when the Labour government whacked up taxes on cigarettes as a post-war fundraising drive. The second was thanks to the failure of Prime Minister Margaret Thatcher's theory of monetarism in the early 1980s—the idea that controlling the level of money in the financial system would keep a check on inflation.

Faced with a rapidly shrinking economy and in need of more cash, Thatcher increased tobacco taxes and the smoking rates fell sharply. Other countries have brought in more strategic tax rises as a way of reducing tobacco consumption, with the French government effectively tripling the price of cigarettes through the 1990s and early 2000s. What followed was a striking example of effective health policy in action: tobacco consumption halved and the government's tax take doubled from 6 billion to 12 billion euros in just a few years. Tobacco taxes work. Governments that are serious about public health and reducing the death toll from cancer just have to be brave enough to pile them on.

## FOREVER YOUNG

So, what should we do to cut our chances of developing cancer? Reducing exposure to the things we know can cause DNA damage is obviously a good idea, as fewer mutations means fewer chances that crucial genes will be hit.

Go to any cancer charity's website and you can read the same prevention advice: don't smoke, take care in the sun, maintain a healthy weight, cut down on booze, eat more fiber and less red meat, be physically active. All of these behaviors are associated with a reduced risk of cancer, but what we don't really know is *why*. It's time to push back against the pseudoscientific bullshit churned out by the wellness industry and start asking serious scientific questions about the impact of different diets, fitness regimens, and supplements on tissue health at a cellular level. Let's stop waffling on about how certain foods or tablets can "boost the immune system," and instead work out whether or not it's possible to manipulate the inflammatory environment in the body and how best to do it safely.

We don't really know what a good tissue micro-environment looks like right now, let alone whether a particular intervention or activity is maintaining it, restoring it, or making it worse. Tomatoes are an interesting case in point. There's some evidence to show that the chemical lycopene in tomatoes can have an anti-inflammatory effect. However, these fruits are members of the nightshade family, which do seem to aggravate inflammation in people who are susceptible.

What we need is better ways of testing these ideas in a rational scientific way, rather than the current system of YouTube videos, blogs, and lucrative book deals based on a ragtag bunch of experiments on cells grown in two-dimensional layers in the lab, animals (enough with the young male mice already!), and exceptionally wealthy individuals with good skin, live-in chefs, and personal trainers. Looking good on Instagram is a poor surrogate for a detailed measure of the tissue micro-environment and its effect on controlling the emergence of cheating cancer cells.

Even so, we'll never be able to avoid the alterations caused by the processes of life at work inside our bodies. Cancer is a bug in the system of life itself: we get cancer because we can't avoid it. DNA sequencing shows that all of our cells—healthy or cancerous—are shot through with mutations, many of which are caused by normal biological processes grinding away at the genome. It's therefore impossible to say exactly what caused any specific tumor. Our bodies have evolved to suppress cancer for a certain length of time, but at some point a cheat will manage to beat the rules once it becomes the fittest cell on its block.

What's clear is that we need a lot more fundamental lab research aimed at understanding what keeps our bodies healthy through our salad days and how we can keep that youthful architecture intact. There's been a lot of focus on what happens once a tumor gets going and growing, but very little work addressing the opposite question: what goes on in healthy tissue to stop cancer from starting in the first place? It's much easier to focus on why we get sick—because that leads to exciting and potentially lucrative research into treatments—and not ask why we stay well.

The fact remains that the single biggest risk factor for cancer is age. But unless you're clever enough to invent a time machine, what can you do about it? It's unhelpful just to smack people in the face with the hourglass of mortality, and "Try not to be old!" doesn't make a great message for a public health campaign. If cancer is an inescapable and inevitable part of life, we really need to think harder about how to prevent it based on a genuine understanding of how the disease starts and spreads.

Persuading every smoker to quit won't reduce cancer cases to zero, although it would prevent millions of premature deaths every year worldwide. Removing all air pollution and toxic industrial chemicals from the environment will only get us so far. Still, however healthily we try to live, none of us are ever getting any younger. For a significant number of people cancer is going to be unavoidable and maybe (if we live long enough) inevitable for us all.

You've probably seen birthday cards with some kind of variation on the slogan "You're as young as you feel!" Well, when it comes to cancer risk, you're as young as your tissue micro-environment. But while TV programs promise the ability to look *10 Years Younger* through good haircuts, plastic surgery, and ever more magical-sounding immortality potions, it's much more important to know what's going on inside.

Similar to Oscar Wilde's eponymous hero Dorian Gray, you might look good on the surface, but the cellular integrity of your internal organs could be the equivalent of a haggard portrait hiding in your biological attic. There needs to be a lot more work done to discover how to keep our tissues young and beautiful in order to suppress the emergence of cellular cheats for as long as possible. Slowing everything down by five or ten years would make a difference; keeping a lid on it for twenty or more would be transformative.

Even the best prevention approaches are only going to get us so far. Another important piece of the picture is diagnosing cancer as early as possible, when swift surgical removal is most likely to provide a cure. Stories about "simple blood tests for cancer" pop up regularly in the news, usually based on the idea that tumors can be detected through the presence of certain faulty molecules or fragments of mutated DNA that have been shed into the bloodstream by dying cancer cells. This technology is very cool and exciting, but the discovery that even healthy tissues can contain "cancer genes" adds an additional layer of complexity. We'll need to be very sure that these kinds of tests are detecting bad tumors that are going to spread and require urgent treatment, and not just flagging up sad clones that are unlikely to cause trouble.

It's also not terribly helpful to have a blood test that reveals the presence of cancer in the body without pinning down exactly where it is. But as imaging techniques like CT scans and X-rays become more sensitive, they're revealing all sorts of lumps and bumps within our bodies. The closer we look—whether for tumors or genetic mutations—the more weirdness we find, just as zooming in on a map of Scotland reveals ever smaller islands, from the obvious ones like Skye, Lewis and Harris, or Mull to tiny specks populated solely by seabirds and seals. How do we know which of the spots on a scan are likely to be dangerous and which are harmless?

By way of example, every year thousands of women who have mammogram breast screening are told the distressing news that they have a lump in their breast known as ductal carcinoma *in situ* (DCIS)—a tiny tumor that may or may not be heading down the road to full-blown cancer. DCIS was virtually unheard of before the widespread adoption of breast screening and now accounts for around a quarter of all breast cancers detected. Some women will have surgery and maybe even chemotherapy and radiotherapy to get rid of it, just to be on the safe side, along with a huge dose of anxiety and stress. But right now we have no way of knowing whether their lump was going to cause a problem or not, adding up to a huge amount of unnecessary treatment and worry.

There's a powerful narrative arguing that because cancer screening saves lives, more screening must be better, right? But it has to be the right screening and it truly has to save lives, rather than bumping up survival statistics by spotting tumors that might never cause a problem. This becomes an even more pressing ethical issue as we understand more about the interplay between cells and the tissue micro-environment. Giving people unnecessary and stressful treatment could end up disrupting their tissue micro-environment, changing the pressures of natural selection and encouraging the emergence of genuinely nasty clones later down the line.

It's tempting to just give up and say, "Well, that's life—nothing you can do about it, might get hit by a bus tomorrow anyway!" But there may be ways to slow down the arrow of time, by finding interventions that maintain small territories of cells rather than letting any single cancerous clone get out of hand. Can we help our tissues to suppress the emergence of cheating cells for as long as possible, whether through lifestyle interventions or pharmaceutical ones? Furthermore, how do we even test these kinds of long-term cancer-prevention measures? It's relatively simple to run a clinical trial for people with cancer whose life expectancy can be measured in months or years, but how do we test protective interventions that might take decades to reveal whether they've worked? One solution might be to come up with some proxy measurements for what's going on in our tissues—for example, by monitoring the shifting patchwork of clones in normal tissue or measuring the range of mutations in DNA shed into the bloodstream. Even so, there are still significant ethical implications of giving a preventative drug to healthy people over many years, which may have unknown long-term side effects.

Through all this detailed work on genetics and clonal evolution, a new view of the origin and development of cancer is starting to emerge. The process of going from a normal cell to an advanced metastatic cancer isn't as simple as picking up a catalog of mutations and we shouldn't expect that treating the disease will be as simple as striking items off the shopping list with particular targeted therapies. The underlying capacity for mutation and selection is present in all of our cells, along with the potential for the emergence of cellular cheats and their subsequent evolution into lethal cancers. And this evolutionary process doesn't stop once a handful of rogue cells have grown into a tumor. It gets worse.

# 6

# SELFISH MONSTERS

The genetic revolution of the nineties and noughties promised to transform the treatment of cancer. By cataloging the rogue genes driving exuberant cell growth and developing highly targeted treatments to block them, researchers and drug companies became confident that cures were in sight. Glivec was the poster child for this march toward precision medicine—a treatment that has truly transformed survival for people with chronic myeloid leukemia (see page 93). Unfortunately, its stellar success has turned out to be extraordinarily difficult to replicate in other cancers.

The example that sticks most in my mind is the drug Zelboraf (vemurafenib). It's a drug designed to shut down the hyperactive version of *BRAF*, a gene fault that drives more than half of all melanoma skin cancers. In 2010, the company behind the drug announced the first results from an early stage clinical trial of PLX4032—the chemical that would eventually become known as Zelboraf. The results were hugely impressive and hit the headlines all over the world. Tumors shrank by at least a third in twenty-four of the thirty-two patients enrolled in the study and disappeared entirely in two more, with very few side effects. Sir Mark Walport, then director of the Wellcome Trust, hailed the breakthrough as a "penicillin moment" for cancer research. In a deeply ironic way, he was absolutely right.

Sitting in a stuffy conference hall a couple of years ago, I gazed in wonder as a researcher flashed up two photos on the screen behind him. The first showed a man in the late stages of melanoma, just before starting Zelboraf. He was gaunt and sallow, tumors sticking out from his frail limbs like knots on twigs. In the second photo the same man was almost unrecognizable: plump and hearty, with no obvious signs of the disease that had been ravaging his body just months before. It was the closest that skeptical scientists ever get to declaring a miracle.

The next slide brought us all back down to earth with an audible gasp. A third photo of the same man, taken a year or so after starting treatment, was almost identical to the first. The knobbly tumors were back with a vengeance, eating his body from the inside out. Although the treatment had gifted him with reasonably good health during the time his cancer was in remission, it had bought a few extra months at best.

This heartbreaking story is all too familiar. Many of us will have known someone with advanced cancer whose disease seemed to be responding well to treatment, then one day it came back and there was nothing more that could be done. The failure of the new generation of molecularly targeted drugs to cure cancer has been both mystifying and disappointing to many in the research community, not to mention patients and their families. The reason why is obvious but has been ignored by the majority of scientists, doctors and pharmaceutical companies for decades. We aren't just trying to treat cancer cells in the body of a patient. We're up against the most fundamental process of life itself: evolution.

The discovery of penicillin in 1928 changed the course of history. Antibiotics turned life-threatening illnesses into minor inconveniences, transformed the safety of surgical procedures, and led to a staggering drop in the number of women dying from infections following childbirth. Whether as rattling bottles of pills or disturbingly artificial-tasting liquids, doctors prescribed antibiotics for everything from sniffles and scrapes to far more serious infections. The discovery that routine antibiotic treatment could improve yields from farm animals led to its widespread use in the USA and many other countries. But humanity's unthinking profligacy with these powerful drugs has had disastrous consequences.

Bacteria can reproduce in just twenty minutes, given the right conditions, going from two to four to eight to millions overnight and picking up mutations in their DNA as they divide. Most mutations are harmful, slowing division or causing cell death, but occasionally an advantageous alteration will arise, such as the ability to grow in the presence of a particular drug. Antibiotic treatment acts as a strong selective pressure, killing off all the bugs that are sensitive to the drug but allowing those that have evolved resistance to thrive. Because bacteria replicate so quickly, this resistant version quickly outgrows the rest and takes over the entire population. Bacteria also like to swap little bits of DNA between themselves, known as plasmids. If resistance to an antibiotic is conferred by a gene lurking on a plasmid, it can be swapped with other bugs very quickly, even if they're a different species.

In less than a century since the discovery of penicillin, the overuse of antibiotics has inexorably led toward the evolution of antimicrobial-resistant superbugs, triggering a global health emergency. As of 2019, around 35,000 people in the US die every year from drug-resistant infections—a number that's set to rise significantly over the coming decade as resistant strains continue to emerge and spread—and doctors are rapidly running out of effective therapies. Experts are warning of a post-antibiotic apocalypse unless rapid steps are taken to come up with better tests, treatments and strategies to combat the evolution of resistance. Frustratingly, this could have been predicted as far back as 1942, when researchers first documented the spread of penicillin-resistant infections.

Cancer isn't a bacterial infection but it's still subject to exactly the same evolutionary pressures as drug-resistant superbugs. A typical cancer already consists of millions of cells by the time of diagnosis, many of which will have their own genetic variations and biological quirks that enable them to adapt and evolve to changes in the world around them—including the appearance of chemotherapy and targeted treatments. Somewhere in there, the genetic seeds of resistance may well have already been sown.

## WELCOME TO THE RESISTANCE

Every year thousands of papers are published in scientific journals, detailing the fruits of the labor of the international community of cancer researchers. Most of them are fairly humdrum, pushing back the frontiers of science a few millimeters at a time. But in March 2012, a paper came out in *The New England Journal of Medicine* that changed everything. It was the brainchild of Professor Charles Swanton at Cancer Research UK's London Research Institute (now part of the Francis Crick Institute)—an ambitious doctor-turned-scientist with access to a useful combination of two things: cancer patients and flashy new DNA-reading machines.

By this point, large studies had already revealed startling genetic diversity between tumors growing in different people, pointing to the need to select targeted therapies based on the genetic makeup of each individual's disease rather than a generic one-size-fits-all treatment. This makes sense given that we now know every cancer arises as an independent evolutionary event with its own unique set of random mutations. But this approach relies on the assumption that all the cells in a tumor are effectively the same, having collected the same set of driver mutations in a straightforward game of genetic bingo. This was partly a limitation of DNA sequencing technology at the time, which needed relatively large quantities of starting material extracted from big tumor chunks or billions of cells grown in the lab, all mashed together in a test tube.

As techniques became more sensitive, things started to look more complicated. In 2006, researchers looking for an altered version of a cancer driver gene called *EGFR*, which makes cells resistant to certain targeted therapies, discovered that a small proportion of cells in lung tumors already carried the resistance mutation before treatment even started. A couple of years later, scientists showed that seemingly identical leukemia cells flowing through the bloodstream could be clearly distinguished into separate tribes based on certain genetic markers.

By 2010, researchers had discovered that secondary tumors that had spread from a primary pancreatic cancer were genetically related to the founder population but appeared to have picked up a bunch of new mutations on their metastatic journey. Then in 2011, a team of Chinese researchers discovered differences in the set of mutated driver genes found in neighboring slices through a single large liver tumor. In the same year, scientists in New York split a small piece of breast tumor down into a hundred single cells and sequenced the DNA in each one. The cells broadly fell into three distinct but related family groups, each with their own unique mix of genetic strengths and weaknesses.

The outline of an unsettling picture was starting to emerge. Every single tumor is actually a patchwork of related but genetically distinct clusters of cells (clones), some of which harbor mutations responsible for metastasis or resistance to treatment. But while all these studies were highly informative, none of them truly captured the genetic diversity within an individual cancer or how these clones had emerged and evolved.

Then came Evie.

Medical confidentiality prevents us from knowing their name or gender, but Evie—Patient EV-001—opened a window onto a world within cancer that had never been seen before. They'd been diagnosed with a large tumor almost taking over one kidney, which had seeded a second one alongside. Evie's lungs were peppered with secondary cancers, while a particularly large growth had set up home in their chest wall. Surgery was the best treatment option, even though the outlook must have looked bleak. But before going under the knife, they decided to volunteer for a clinical trial testing whether or not a six-week course of the drug Afinitor (everolimus) could shrink the tumors and make them easier to remove, continuing the treatment afterwards if it seemed to be helping.

As Evie was undergoing surgery, Swanton and his team collected the tumors and chopped them into pieces—nine from the large primary, two from the one in their chest and the whole of the little secondary kidney tumor for good measure. Then they spent three years painstakingly analyzing the DNA from each of them, putting together a catalog of all the genetic changes they could find. The results were fascinating and confounding: although all the samples were clearly related and shared a number of mutations in common, no two were genetically identical—not even clumps sitting right next to each other. What's more, the distant secondary tumors were remarkably different from the primary cancers they sprang from. The next step was to work out how all these clusters of cells were related, in order to map out the evolutionary journey they'd been on.

They did it like this. Imagine you're looking at a photo of all the people in a very large, very strange family from a faraway land. First, you notice that every single one of them has bright blue hair, whereas all the other people in the country are dark-haired. That tells you that the gene change responsible for blue hair must have happened a long time ago and was probably the first thing to distinguish this unusual bunch from more regular folk.

Then you spot that about half of the family members have six fingers on each hand, while the rest have five. This genetic alteration must have happened after the hair color, yet still early on when there were few people in the family—one half got the six-finger mutation, so all of their descendants that inherit it will also carry it—while the other didn't.

Finally, you realize that each person has different-colored eyes—red, yellow, green, purple, and more, as well as all sorts of other unique traits. This last set of gene changes must have arisen very recently in each person, as they're specific to each individual rather than shared among the whole group.

This is enough information for an amateur genealogist to build a simple tree showing how the family must have split and changed over time, as well as figuring out the relationships of the underlying genetic faults: the hair gene changed first in the clan's ancestral founder, then the gene for finger number, then eye color and everything else. Applying the same principle to the genetic data from all the little chunks from Evie's tumors, Swanton was able to piece together a family tree for the different cell clones, identifying each new genetic change splitting off as a branch from the original trunk. Multiple samples from a further three patients from the trial confirmed that what they were seeing was true: every tumor was made up of related but distinct clones, each carrying shared and unique driver mutations.

The neatly traced simple family trees in the resulting paper in *The New England Journal of Medicine* look unnervingly familiar to one drawn almost two centuries earlier by another scientist. In a beautiful moment of scientific synchronicity, he was also called Charles.

One day in 1837, barnacle obsessive and bassoon enthusiast* Charles Darwin turned a fresh page in his notebook and wrote the words "I think." Beneath that he sketched out his idea for a tree of life, with new species branching away from older extinct specimens as they adapted and changed over time. This simple concept lies at the heart of his theory of evolution by natural selection, finally published in 1859 after a lengthy period of procrastination. Evidence from every strand of science from geology to genetics has since supported his proposal that evolution underpins the diversity of life on Earth.

Evolution has been at work on this planet for 4 billion

* Darwin had been fascinated by earthworms for many decades of his life and his final book, published six months before he died, was a treatise on their behavior. To test whether or not worms could hear, he set about making what must have been an almighty racket—Darwin playing a tin whistle and his son parping loudly on a bassoon, followed by loud shouts and thumping of piano keys—eventually concluding that while they were sensitive to vibrations in the air, worms were immune to the Darwin family's musical talents.

years, shaping organisms to suit environments ranging from murky oceans to breath-taking mountaintops. Random genetic changes (usually the result of blips and slips in DNA copying and cell division, or the impact of external forces such as radiation or chemicals) led to species with slightly different characteristics. Most of these changes will be either harmful or have no impact, but a tiny handful will be a lucky biological bonus—a tweak that makes the bearer slightly bigger, stronger, smaller, smarter, sturdier, stripier, or spottier than the rest of its kin.

This works to its advantage in the face of predation, food shortages, lack of space, changing climate, or any other selective pressure you can imagine. As a result, these ever so slightly souped-up animals, plants, or bugs are ever so slightly more likely to reproduce and pass on their useful genetic arsenal to the next generation. Rinse and repeat over millions of years and here we are today, with a planet covered by a patchwork of more or less distantly related species, each able to trace their genetic roots back to a 4-billion-year-old common ancestor, LUCA.

Just as Charles Darwin's conclusions about the origin of species were ultimately inescapable—organisms adapt and change in response to selective pressures—Charles Swanton's results tell us that cancers in the body behave in the same way. A large population of genetically messed-up, rapidly replicating cancer cells is a microcosm of evolution with each little pocket of cells going off on its own *Choose Your Own Adventure* story. Secondary tumors are more distant relatives, with their own set of molecular quirks and foibles. All of these clones came from the same single founder cell and diverged as the disease developed, picking up new mutations and alterations along the way.

Before you get the impression that cancer is a disease solely of genes and genetics, there's also epigenetics—the "nurture" bit of the Nature Plus Nurture equation. Our genome is plastered with all sorts of molecular marks and tags, known as epigenetic modifications, which can't easily be detected using simple DNA sequencing techniques. Like sticky notes placed in a recipe book to flag up favorite recipes, these modifications mark out patterns of gene activity in response to changes in the environment both within and outside the body, including diet, stress, exercise, and all the rest. Many of these modifications are messed up in tumors and are likely to play a role in enabling cancer cells to adapt to changes in their local environment by switching genes on or off without necessarily having to pick up a mutation. As an example, a vital DNA repair gene called *MLH1* is switched off in some bowel cancers in response to low oxygen levels—a change that's only detectable by specifically looking for alterations in epigenetic marks around the gene rather than any underlying mutation that could be spotted through DNA sequencing.

Any technique for looking for the mutations in a tumor that relies solely on mashing up large chunks of tissue will miss all this exquisite diversity, like blending up a smoothie made of twenty types of whole fruits and expecting to be able to discern the taste of a single blueberry against an entire pineapple. This is particularly relevant when it comes to small clumps of cells harboring mutations that make them resistant to treatment, which may not be obvious at first but turn out to be lethal further down the line.

Scientists refer to this mutational patchwork as "tumor heterogeneity" and it can tell us a lot about how an individual cancer has grown and changed over time at the basic genetic level. On a global scale, evolution has led to the glorious diversity of species on Earth. But genetic diversity within tumors is a big problem and it's the reason why most treatments for advanced cancer ultimately fail. In nature, genetic variation among organisms within a species means that there's usually a few hardy specimens that can adapt and survive even in the face of the harshest conditions.

In cancer, the onslaught of radiotherapy, chemotherapy, or molecularly targeted drugs acts as a selective pressure, weeding out sensitive cells and killing them. Yet there are likely to be a few pockets of resistant cells that make it through to the other side and start growing again. It's not the fault of the treatment, it's just evolution in action. The same feature that creates thriving biodiversity on Earth is also a bug in the system of life. And like a comic book villain emerging from a toxic swamp with ten times the strength and twice the brutality, what doesn't kill cancer only makes it stronger.

## TREES AND TRUNKS

It's easier to understand cancer as an evolutionary process if we expand our view from the claustrophobic confines of the tissue micro-environment into the wider world. Imagine an ecologist trying to measure the biodiversity in a large area of African savannah—sturdy boots, stained shorts, beard optional. It's simply impractical to survey every inch of it, so instead they'll pick out a few smaller areas, known as quadrants, and count the animals and plants in each, hoping that it gives them a representative sample of the whole. However, this approach risks missing out on rare but ecologically important species that happen to fall outside the survey areas. Similarly, genetic analysis of a tumor tends to look at a single snapshot in the hope that it represents the disease but misses rare cellular "species" that could be carrying the seeds of resistance to treatment.

Since those early days, many researchers around the world have started studying genetic heterogeneity in all kinds of tumors, cataloging the evolutionary changes in related clumps of cells and trying to map that on to how fast the cancer grows and becomes resistant to therapy. The family trees of some tumors look like date palms, with very little genetic change for a long time then a short burst of evolutionary activity. These are the kinds of cancers that might be cured if it's possible to find a drug that hits a mutation that's present in the "trunk" and shared by all the cells. Others are more like sturdy oaks, with several distinct branches spreading out as new driver mutations drive clones of cells to expand. Using a combination of several drugs might provide a way to prune the branches, but it's not as simple as slashing through a palm tree. And some are like rampant shrubs, with multiple branches rushing straight out of the ground and entwining all over the place, evading any attempts at chemotherapy.*

Just as Charles Swanton and his team were adding the final touches to their 2012 kidney cancer paper, a bright new graduate student called Nicholas McGranahan joined the lab. A few impossibly short years later, that student became a group leader at University College London (UCL). Still working closely with Swanton, McGranahan is now focusing on lung cancer—the leading cause of cancer death in the world.

Although many lung tumors are found at a relatively early stage, when they're still quite small, they tend to spread prodigiously and aggressively. Surgery and radiotherapy aren't always possible, and chemo is pretty much ineffective. Despite the development of a number of targeted drugs aimed at the most common genetic changes in lung tumors, resistance is almost guaranteed and survival is usually measured in months rather than years. As part of a major project known as TRACERx (Tracking Cancer Evolution through treatment (Rx)), Swanton, McGranahan, and their colleagues are mapping out the genetic patchwork in more than 800 patients, taking multiple samples at every step from diagnosis through treatment and relapse.

Frustratingly, their findings are raising more questions than answers. Even the smallest cancers are miniature patchwork blankets stitched together from even smaller clones of

* Ironically, these highly mutated, branching tumors appear to be better candidates for new immunotherapy drugs, as their flamboyant genetic weirdness is more likely to attract the attention of the immune system than plainer, more stealthy tumors.

genetically different cells. To make things even more confusing, all too often the same mutation turns up in two or more regions of a tumor, giving the impression that it must be universal (or "clonal," to use the scientific term), only for another patch of cancer cells to be found without it.

This "illusion of clonality," as McGranahan describes it, is particularly prevalent in lung cancer and it's a problem when it comes to selecting the most appropriate targeted therapy for that patient. If at all possible, you want to pick a drug that targets the product of one of these universal clonal mutations, so it's got a good chance of hitting every cancer cell in the body. But if it turns out that you've accidentally missed out on sequencing a pocket of cells *without* that mutation, then the treatment will inevitably fail sooner or later. But wait—it gets worse.

To add to the confusion, McGranahan explains that not only can cancer cells pick up new mutations as they evolve, but they can also sometimes manage to fix themselves. He and his colleagues have found an increasing number of cases in which all the cells in a tumor carry a particular mutation early on in the disease. Yet, when they look at a later point, they find descendants of these damaged cells that appear to have healed themselves. This could be significant if the repair provides a selective advantage, such as making them resistant to treatment. The ability of cancer cells both to break and fix themselves highlights just how dynamic these evolutionary processes are.

What's more, although cancer cells can seem like a wily foe, twisting and turning at every opportunity and exploiting new niches to survive, this evolutionary flexibility comes at a cost. Evolution is not an artisan or engineer, elegantly designing the optimal solution to any problem. It's more like MacGyver—the eponymous 1980s American TV character—grabbing the genetic stuff it has at hand to get out of a sticky situation and survive. Unlike our hero, who always got out of his scrapes before the credits rolled, the vast majority of cancer cells don't make it out alive. Most mutations are harmful under normal circumstances, leading to cell death or serious slowdown. It's only during the harsh selective onslaught of treatment that they can save the day.

The kinds of changes that render cells resistant to therapy also slow them down, meaning that most of the time they don't grow and multiply as fast as drug-sensitive cells. But if the more vigorous cells are killed off by a bout of chemotherapy or targeted treatment, suddenly the slow-growers are the only ones left. So it's not enough just to target cells that seem to be growing and thriving—it's the quiet ones that we need to watch too.

Depressingly, scientists have discovered that the seeds of resistance to targeted therapies can be there right from the very start. One detailed study of a patient with myeloma—a cancer affecting white blood cells—revealed that the cancer cells that finally overwhelmed them grew out of a tiny clone that was present at the very earliest stages of their disease. Yet, somehow it resisted all the treatments that doctors could throw at it, eventually outcompeting and overpowering everything else.

Another paper in 2016 comparing the genetic makeup of

samples from medulloblastoma brain tumors taken before and after treatment shows the stark reality of natural selection at work. In more than thirty patients, researchers found that the cancer cells that grew back after treatment had been present in the original tumor, but only in a tiny pocket. Once the bulk of the tumor was killed off by radiotherapy, these inconspicuous resistant cells had the run of the place and could quickly take over. In many cases none of the original cell clones containing what were thought to be ubiquitous driver mutations were still there after treatment. If this were a gangster movie, it would be like the unassuming apprentice in the corner rising up to become the criminal kingpin once all the competition has been eliminated.

There's another problem, and it's so rarely mentioned that it's become a kind of dirty secret in the world of cancer therapy: many conventional chemotherapy drugs and radiotherapy work by damaging DNA, creating even more mutations that could be contributing to resistance. Newer targeted therapies aren't exempt from blame either. A study published at the end of 2019 showed that bowel cancer cells activated more error-prone DNA copying mechanisms in response to drugs targeting specific molecular alterations, increasing the chances of introducing new mutations that allow them to evolve resistance to therapy.

In 2012, scientists at Washington University in Missouri looked at DNA from eight patients with acute myeloid leukemia (AML) who'd all been given chemotherapy yet relapsed within a couple of years. In all eight cases, they found evidence of genetic changes that seemed to be the handiwork of the treatment. In their paper, published in the journal *Nature*, the researchers note that: "Although chemotherapy is required to induce initial remissions in AML patients, our data also raise the possibility that it contributes to relapse by generating new mutations."

In other words, although chemotherapy is the only treatment that works to treat leukemia, it might ultimately end up making things worse for some patients. However, this doesn't seem to be the case all the time. Researchers have looked at a handful of brain tumor patients who'd been treated with the drug Temodal (temozolomide), which causes a specific and highly recognizable type of DNA damage. They found that the DNA of the surviving cancer cells was peppered with new mutations that were unmistakably the work of the drug. But they didn't see it in every person and no one's sure what lies behind the difference. Whatever the reason, it's definitely Not Good.

## DARWIN'S REVENGE

As DNA sequencing continues to get faster, cheaper, and more sensitive, researchers around the world have started to unpick the genetic patchwork of cancer in more detail. Results have started to come in thick and fast, revealing complex patterns of genetically distinct cells in esophageal, ovarian, and bowel tumors, and many more. In fact, the genetic fingerprint of one group of cancer cells may more closely resemble that from a sample taken from a completely different patient than a cluster of cells growing right beside them.

Any of these tiny clones could harbor genetic changes that render them resistant to treatment—even the newest, most expensive targeted therapies—and it may only take a single escapee cell to start a tumor growing again.

It's a truth that few doctors and researchers—and even fewer patients—want to hear. Tumors evolve and diversify as they grow, and even a tiny pocket of resistant cells is enough to bring the disease roaring back. Charles Swanton's *New England Journal* paper put a rocket up the cancer research community, making tumor evolution and heterogeneity the hottest of hot topics and triggering major research programs around the world. But it really shouldn't have come as that much of a surprise.

More than forty years ago, Philadelphia-born scientist Peter Nowell wrote a short article in the prestigious journal *Science* entitled "The clonal evolution of tumor cell populations." He argued that while cancers start from a single cell, they evolve through rounds of mutation and selection to become more aggressive and resistant to treatment. He even saw the need for personally tailored treatment according to the genetic makeup of each individual patient's disease, foreshadowing the era of precision medicine by a good few decades. With unnerving prescience, the last two lines of the paper's summary read:

> Hence, each patient's cancer may require individual specific therapy, and even this may be thwarted by emergence of a genetically variant subline resistant to the treatment. More research should be

> directed toward understanding and controlling the evolutionary process in tumors before it reaches the late stage usually seen in clinical cancer.

By the time Nowell's paper was published in 1976, he'd already made a name for himself as the co-discoverer of the so-called Philadelphia chromosome (page 92)—the stubby fragment of DNA responsible for driving the runaway growth of the blood cancer chronic myeloid leukemia. Despite his fame (at least in the world of science), few people took much notice. One who did was Professor Mel Greaves, a childhood leukemia specialist now based at The Institute of Cancer Research (ICR) in Surrey.

More than twenty years ago—before we knew anything about the extent of the genetic heterogeneity in tumors—he published a book entitled *Cancer: The Evolutionary Legacy*. In it, he outlined the idea of cancer as intrinsically and inescapably tied to evolution, both of our own human species and the disease itself as it emerges and spreads within the body. Even without the hindsight of modern genomics, it still bears up as a devastatingly clear explanation of the principles of natural selection at work within the genetically distinct clusters of cells in a tumor. But, like Nowell before him, nobody paid much attention.

According to an analysis published in 2011, only around 1 percent of all scientific papers about cancer relapse or resistance to therapy published since the 1980s mentioned the concept of evolution, rising to a still-paltry 10 percent or so over the following five years. To some extent this is understandable. It's only in the past few years that DNA sequencing technology has advanced to the point where it's feasible to read all the genes in a relatively small tumor sample and trace its evolutionary journey, let alone repeat that tens or even thousands of times. Nowadays, there's no excuse.

In search of an explanation for why the evolutionary view of cancer was ignored for so long, I took my well-thumbed copy of *Cancer: The Evolutionary Legacy* for a trip to the ICR to meet its maker. Now approaching eighty and the recent recipient of a knighthood, Mel Greaves was a postdoctoral researcher who'd just started working on cancer when Nowell's *Science* paper came out. His background in evolutionary biology, taught him by the great mathematician and geneticist John Maynard Smith, made him quick to grasp the significance of Nowell's idea when so few others in the field took it seriously.

"It still surprises me how little effect that paper had, but I thought it made absolute sense—it's just the principle of how biology works, so it was really surprising to me that people didn't get it," he says, his quiet voice rising in irritation. "Darwin worked all of this out without knowing anything about DNA or genes! It's so rivetingly simple, it just grabs you—so why isn't everybody getting onto it?"

I think back to the example of the Vogelgram—the neat linear diagram showing the progression from cell to lump to tumor as the mutational bingo card fills up—which has had a huge influence on the way that researchers and doctors think about the development of cancer. In many ways this is a lot like the iconic "March of Progress," created in 1965 by the artist Rudolph Zallinger, showing human evolution as a straightforward sequence from knuckle-dragging monkey to lumbering ape to thick-browed cave dweller to a pleasingly ripped, upstanding modern man.

It's a visually arresting and elegant image that was never intended as a scientifically accurate depiction of the process, yet it's been inadvertently misdirecting the public perception of evolution ever since. Even back in the 1960s, it was starting to become clear that species evolve gradually over millions of years, with plenty of dead ends and diversions along the way, creating evolutionary trees that are knotted, tangled shrubberies rather than linear trajectories.

*Homo sapiens* is the only human species alive today, having seen off all our ancestral hominin competitors. Yet there's plenty of evidence of their existence in the fossil record, along with intriguing results from analysis of their ancient DNA, showing that there was a hell of a lot of inter-species hanky-panky along the way. We've also continued to evolve alongside the descendants of our more distant relatives—monkeys, chimps, gorillas, bonobos, and all the rest. Trying to explain the complex evolutionary relationships of all the primates on Earth with the "March of Progress" is as futile as trying to explain the complexity of an advanced metastatic cancer with the Vogelgram and a handful of driver mutations.

"Genetics and genomics are fantastic—we can see all this complicated variation and it's not just a sequence of events—but I still think we are far too gene-centric," Greaves argues. "I'm pushing a more evolutionary model because it's the context in which things are happening, and it astonishes me that more oncologists aren't aware of what drug resistance is. It's just Darwinian selection, for goodness' sake, but it's taking a long time to sink in!"

To people like Mel Greaves and Charles Swanton, the evolutionary nature of cancer may seem blindingly obvious. Others have found it hard to wrap their heads around the idea that cancer is a constantly changing, complex system that adapts and evolves rather than a fixed target that can be eradicated with a single shot. Perhaps there's too much invested by large funding organizations and the pharmaceutical industry in the hunt for the Magic Bullet to admit that this idea was always destined to be thwarted by Darwinism.

In fact, as Greaves points out to me, the more precisely honed a therapy is to a single molecular target, the faster we can expect resistance to emerge. By analogy, most animals can quickly adapt to the loss of a single food source in their environment by feeding on something else, unless it happens to be the only thing they can eat.

This also helps to explain the unique and unrepeatable success of Glivec in treating chronic myeloid leukemia—a disease driven by a single driver gene created from the fused Philadelphia chromosome (page 92), which is found in every single cancer cell. And because CML is so completely dependent on this fusion gene for its survival, knocking it out effectively eliminates it altogether. Glivec is the closest thing to a Magic Bullet that we have, but it fooled us all into thinking that curing cancer was going to be a straightforward process of finding more of the same.

## THE CRUCIBLE OF EVOLUTION

Half a billion years ago, life was simple. Bacteria, amoebas, maybe a few little multicellular animals. Then everything changed. Over a brief period of 70 to 80 million years,* life on Earth went through an accelerated period of evolution known as the Cambrian explosion, during which most of the major groups of animals first emerged.

What emerged from deep in the soupy waters of the giant primeval ocean was a collection of creatures so improbable that they'd be rejected from a horror movie's props department for being too ridiculous. One of them, more formally known as *Odontogriphus*, looks like a long flappy Roomba, while *Wiwaxia* appears to be a Viking helmet with frond-like fingers instead of horns. *Anomalocaris* is what you'd get if you crossed a lobster with a can opener, *Nectocaris* is a boggle-eyed squid with two waving tentacles and *Opabinia* looks like a five-eyed prawn that swallowed a vacuum cleaner. Then there's the appropriately named *Hallucigenia*—a thumb-sized worm bristling with legs, teeth, and spikes, fresh from your worst nightmare.

These bizarre animals are "hopeful monsters"—the term used by German biologist Richard Goldschmidt to describe their fossilized remains in the Burgess Shale, a strip of prehistoric sea floor that ended up sandwiched within the Canadian Rockies. The incredible inhabitants of the Burgess Shale were unlike anything that had come before or since.

* Brief in geological terms, which explains why paleontologists are always late.

The underlying trigger for the Cambrian explosion remains a mystery—if, indeed, there was one sole cause. Researchers have suggested all kinds of explanations ranging from a sudden increase in oxygen levels, flooding of low-lying land masses leaching nutrients into the sea, or even cosmic radiation bursts from the Milky Way. There may also have been a few chance innovations that kick-started the rapid diversification of species, such as the ability to swim into higher waters or burrow through the slimy mat of bacteria that lay over the ocean floor like a layer of cling film, revealing new niches and food sources to explore. Another possibility is the evolution of vision, setting off a biological arms race between newly sighted predators and their prey to eat and avoid being eaten—not so much a dog-eat-dog world as a monster-eat-monster one.

Whatever the spark, the Cambrian explosion was fundamentally a genetic boom. Instead of a slow, creeping process of natural selection fueled by small mutations, evolution got turned up to eleven. The exuberant experimental rearrangement of structures in Goldschmidt's hopeful monsters was most likely the result of unprecedented shuffling within their genes as these strange creatures battled it out for survival in the ocean. Yet, despite the emergence of endless forms most wonderful in the primeval sea, virtually none of them were successful. Most went extinct, leaving no trace of their existence in today's world except a flattened imprint in a Canadian mountainside. But a few of them survived, evolving into more conventional-looking organisms by our modern biological beauty standards.

We often have an idea of evolution as being a stately change that takes place over eons, driven by small, stepwise genetic changes. It took 50 million years for a dog-sized, four-legged, four-toed mammal, *Hyracotherium*, to evolve into modern horses—a slightly longer neck here, a slightly sturdier hoof there. If you met a *Hyracotherium* in the woods, you'd have to squint very hard to see the inklings of its horsey descendants, but at least it's in the right kind of animal ballpark. This kind of gradual evolutionary journey is similar to what's seen in early stage cancers or childhood tumors. But the wild genetic diversity in advanced cancers looks more like the profusion of weird characters in the Burgess Shale.

Echoing Goldschmidt's sentiments about the creatures of the Cambrian ocean, cancers are "selfish monsters," evolving wildly and rapidly in the space of a patient's lifetime. The same fundamental biological processes are still at work, sped up to breakneck pace rather than glacially sliding through deep time. Given enough cells, time, genetic fuel, and selection pressure, just about anything you can imagine will happen inside the crucible of cancer, with new cells being born, dying, and shuffling their genomes in a race to adapt or die. The winners are those that land on innovations enabling them to survive and proliferate another day, while the losers go extinct.

Thanks to the team at the Sanger Institute, we now know that by middle age our bodies are a patchwork of mutated cell clones, jostling for space in the crowded tissue environment. Most of those mutations are relatively small, specific changes, some of which happen to be in so-called cancer driver genes. Other than that, the rest of the genome looks fairly normal. Although it's currently impossible to pinpoint the exact moment a growing cluster of dubious cells tips into becoming a cancer, a key event in many cases seems to be the onset of the kind of genetic shuffling that fueled the Cambrian explosion, known as chromosomal instability. On top of the gradual tick of minor mutations, the genomes of cancer cells are a mess. Whole stretches of genes are copied or deleted, large chunks of DNA appear to have gone on walkies throughout the genome, while entire chromosomes are duplicated or missing entirely.

Using DNA sequencing data to reconstruct the evolutionary trajectories of individual cancers, researchers have discovered that precancerous lumps and slow-growing tumors tend to follow a more sedate path of Darwinian evolution, gradually picking up a few new driver mutations here and there and progressing over a decade or more. Others show signs of chromosomal instability very early on, leading to rapid growth, speedy spreading throughout the body and a high chance of being resistant to therapy. A genetic Big Bang lights the fire under the evolutionary crucible within a tumor. And once that happens, it's very hard to stop.

We've known that cancer cells often have the wrong number of chromosomes since the days of Hansemann and Boveri more than a century ago. This phenomenon, known as aneuploidy, happens when chromosomes aren't correctly allocated during cell division (mitosis). One daughter ends up with too many, while the other gets too few, ending up with cells that are either missing or getting a double dose of thousands of genes at once. Sometimes the cell just skips the entire division step and does another round of DNA replication, doubling the entire genome.

This might not matter so much if the number of chromosomes (and therefore genes) is still evenly balanced, but having that many chromosomes makes further divisions tricky, increasing the chances that one or more chromosomes will go missing next time around the cell cycle. And having extra DNA means more genetic fuel for evolution, as any of these bonus chromosomes is likely to pick up further mutations of its own.

Imagine playing soccer with twenty-two players on each side instead of eleven—the game will probably be quite confusing, but it might just about work out. Removing all the forwards from one team or taking away the goalkeeper creates a dramatic imbalance between the teams, making it much more likely that one will win. And as each of those players starts to pick up their own collection of injuries—the equivalent of further mutations in a cancer—then the match becomes increasingly unbalanced.

This question of chromosome imbalance is something that's always intrigued Rong Li. Today, she's Professor of Cell Biology at The Johns Hopkins University School of Medicine in Baltimore, but the fascination with mitosis has been there since her early days as a graduate student in the 1980s. This was a big time for cell division research, as the molecular engine that powers the cell cycle had just been discovered, opening up a rich new seam of research opportunities. As a budding young group leader, Li turned her attention to studying mitosis in yeast, hoping to understand how chromosomes always ended up being correctly allocated into each daughter cell. Despite her diligent planning and careful experiments, it was an unexpected result that turned out to be one of the most important in her early career.

Li and her colleagues were using genetic engineering to remove a gene called Myosin II from yeast. This encodes a tiny molecular motor that creates the physical forces driving the final separation phase of cell division. It's almost identical in all organisms from yeast to humans, suggesting that it's pretty fundamental to life, so Li's assumption was that the cells would die without it. She was wrong.

Although most of the modified yeast cells died, there were a few hardy survivors—strange-looking cells clinging to each other and clearly struggling to divide. Out of curiosity, Li carefully picked them up and put them in a dish of fresh food to see what would happen. To her surprise, a few tiny colonies started to grow. Again, she picked out the biggest and put them in a new dish. This time, there were more colonies and they were growing faster. Once more, she picked the best and transferred them to a new home, and kept on repeating the cycle.

Li was effectively managing the ultimate yeast knockout challenge, with only the toughest survivors going through to the next round. After more than ten rounds of selection, she ended up with colonies of yeast that were indistinguishable from normal cells yet were still missing what had previously been considered to be an absolutely vital gene. They'd evolved in front of her eyes to overcome a major genetic roadblock, but when she looked to see whether they'd picked up further mutations that might somehow be compensating for the loss, she drew a blank. So how on Earth were they doing it?

The only thing she could find wrong was aneuploidy—all the yeast had different numbers of chromosomes, instead of the usual sixteen. Not only had they learned to live with this unusual arrangement, but they were also using their extra chromosomes as a tool to create new traits. Out of forty-five strains of super-yeast that she managed to cultivate, ten were growing as well as normal cells. But they'd only managed to evolve three ways to solve the problem of having no Myosin II, with common behaviors and patterns of chromosome gain or loss. One solution is to build a very thick cell wall at the point where the cell is supposed to divide, physically squishing it in half rather than using the motor to drive the split. All of the superstrains that had evolved this solution turned out to have an extra copy of chromosome 16, which carries a couple of genes that are essential for cell wall production. A double dose of these genes means a double dose of cell wall, removing the need for the myosin motor.

It might sound like a useful cellular superpower, but having extra copies of chromosomes or genes isn't usually an advantage if these cells are competing against cells with a regular number of chromosomes in a normal environment. Aneuploidy is effectively a massive mutation, altering the activity of hundreds or thousands of genes in one fell swoop. But once the environment changes then the rules of the game change, too.

As long as cells are happy and proliferating in a good environment, aneuploidy is a disadvantage. If aneuploid yeast cells grow side by side with regular ones, then they're quickly outcompeted by their normal neighbors. But when the cells get stressed, whether through genetic mutations in important genes like Myosin II or unpleasant environmental conditions, survival becomes the overriding driving force. Aneuploidy is a drastic solution under normal circumstances, but it quickly generates an enormous number of potential genetic permutations and combinations, providing rich fuel for evolving out of a sticky situation. In the vast majority of circumstances, the outcome is likely to be bad—the usual outcome of any chromosomal imbalance is cell death or a permanent block on division—but very occasionally something useful will emerge out of the chaos. Not so much survival of the fittest as survival of the weirdest.

This response goes all the way back to bacteria, which activate a sloppy, error-prone DNA repair toolkit when times get tough as a way of increasing the diversity in their genome to evolve their way out of trouble. And although the exact details aren't known, it looks like cells in more complex organisms—including humans—will loosen the quality controls on cell division that normally ensure that all of their chromosomes are copied and separated correctly, shuffling the genetic deck in the hope of coming up with a survival plan.

Yeast has sixteen chromosomes that can vary from one to four copies in aneuploid cells, even without adding in the impact of further rearrangements and mutations. So just imagine how much more variation might be accessible by shuffling the 46 chromosomes making up the human genome in a cancer cell. It's not surprising that the vast majority of human tumors are aneuploid and the more advanced they are, the more weird their chromosomes are. Chromosomal instability seems to be a fundamental property of the most aggressive and hard-to-treat tumors, with unstable cancer cells gaining or losing chromosomes every five divisions compared with more stable cells that only screw it up once in a hundred times.

This quickly becomes a vicious circle: stressed cells are more likely to have wonky cell divisions that generate aneuploidy, leading to disrupted levels of gene activity and unbalanced numbers of chromosomes, which makes the cells more stressed and prone to making even more mistakes. Some cancer driver mutations also speed up the cell cycle, increasing the likelihood that the process will be rushed through without all the correct checks being made. Several research teams are now investigating whether or not drugs that target parts of the molecular machinery responsible for turning the wheels of the cell cycle could help to slow things down, giving cells time to check their work and making it less likely that chromosomes will go astray. It's an intriguing idea that might stop nascent tumors from tipping into the instability danger zone in the first place.

Another trick used by cancer cells is known as genome doubling, where they duplicate their entire genome. This happens when a cell has copied all of its DNA in readiness to divide but fails to follow through on the deal. In some cases this can be more than fifteen or twenty years before a cancer is finally detected. In one stroke, a cell has twice as much DNA to play with, which makes potent fodder for evolution. Extra copies of genes can act as "backups" in case the originals get broken, or they may mutate into new malevolent forms. This doubled capacity for genetic experimentation is an evolutionary shortcut to the creation of unusual cells. It's a tactic that's also been put to use in the history of agriculture—many common fruits, vegetables, and grains have multiple sets of DNA, resulting from natural genome doublings throwing up interesting variations that are then selected by farmers for further propagation.*

Aneuploidy and genome doubling aren't the only way to cause chromosomal chaos in cancer cells, although they're extremely common. There's a multitude of other ways that tumors end up with messed-up genomes. The kind of stresses encountered by cancer cells can reactivate some of the millions of dormant virus-like sequences that lurk within our DNA. Known as transposons, these rejuvenated zombie genes start jumping around the genome, dragging nearby genes with them as they go. There are multiple acts of genetic vandalism, where large or small sections of chromosomes get cut and pasted into new locations, often the result of DNA repair processes gone awry.

* Many cancer cells have twice the usual two sets of human chromosomes, making them tetraploid (regular cells are referred to as diploid, while eggs and sperm are haploid). However, the humble strawberry can have two, four, five, six, seven, eight, or ten copies of the seven basic strawberry chromosomes, making it di-, tetra-, penta-, hexa-, hepta-, octo-, or even decaploid.

Telomeres—the protective caps on the ends of chromosomes, which gradually get shorter every time a cell divides (see page 30)—can wear down to nothing if a cell keeps on multiplying for too long. These frayed ends look a lot like the kind of damage seen when DNA snaps in two, fooling the cell's repair machinery into fixing the breaks by gluing whole chromosomes together at the ends.

Then there's chromothripsis—a wonderful name derived from the Greek word meaning "shattered into pieces." In 2011, a team of researchers who were sequencing cancer genomes discovered that in a small but significant proportion of all tumors, entire chromosomes had been broken into pieces and glued back together again. Rather than a neat repair job, these chromosomes are patched back together in a haphazard and chaotic manner, like a smashed stained glass window put back together with no attention paid to preserving the original design. Chromothripsis is particularly common in certain types of cancer, including up to a quarter of bone tumors, and is a potent source of evolutionary fuel.

Just like the chromosome fusion that brings the *BCR* and *ABL* genes together in the Philadelphia chromosome in leukemia, these fusions slam together genes that should never be neighbors to create a lethal cancer-driving force. In other cases, chromosome rearrangements reorganize the genetic filing system within the nucleus, moving normally active tumor suppressor genes into areas that are designated as silent, or bringing inactivated drivers into an activity zone.

Using colorful microscopy techniques that highlight each

chromosome with a different fluorescent dye, modern-day Boveris and Hansemanns have seen all kinds of weird and wonderful genetic rearrangements in tumors. Chromosomes are copied or missing altogether, completely shattered and stitched back together in new conformations, or even joined into strange little rings packed with highly active cancer-driving oncogenes. Down at the level of each individual cell these events seem like a catastrophic last-ditch panic option with a minimal chance of survival. But when viewing a collection of cancer cells as a population of rapidly diversifying miniature organisms within an environment, it's easy to see the evolutionary imperative at work. Just a handful of survivors are enough to ensure the future of the group. It doesn't matter how weird they are—all that matters is that they make it.

There are more than seventy individual chromosomes in HeLa cells—Henrietta Lacks' cervical cancer cells that have been grown in labs all over the world since the 1950s*—along with more than a million and a half single-letter DNA typos, nearly 750 large and 15,000 small sections of chromosomes missing, and the best part of 3,500 new sections pasted back in. The evolutionary biologist Leigh Van Valen argued that HeLa's long life in the laboratory and significant divergence from the normal chromosomal arrangement of human cells warranted its designation as an entirely new

* Read more about the fascinating and important story of Henrietta and her cells in *The Immortal Life of Henrietta Lacks* by Rebecca Skloot (Crown Publishing Group, 2010).

species, *Helacyton gartleri*, although few other scientists support this suggestion.*

However, in the natural world, entirely new species emerge when there are significant chromosomal differences between formerly closely-related populations. For example, humans share more than 98 percent of our genome with chimpanzees and the main thing that distinguishes the two species at a chromosomal level is a single rearrangement. At some point in our ape ancestry, two chromosomes fused together to create what's now human chromosome 2, leaving us with twenty-three pairs of chromosomes while chimps still have twenty-four. Given the much more dramatic chromosomal chaos within the cells roaming the dystopian landscape within a late-stage tumor, it could even be argued that each person's cancer also represents one or more entirely new species.

## SPINNING THE WHEEL OF DEATH

In the past few years we've got really good at reading genes. Endless strings of A, C, T, and G are spewing out of high-throughput gene machines all over the world, revealing the genetic makeup (genotype) of thousands of tumors and

* The name is a nod to molecular biologist Stanley Gartler, who discovered that a huge number of what were thought to be cancer cell lines growing in labs all over the world were actually HeLa in disguise. This cross-contamination has been known about for decades and is still going on today, yet there's an unsurprising reluctance on the part of researchers to admit that the contents of many *in vitro* cell experiments may not be quite what they say on the lid of the Petri dish or acknowledge the implications for cancer research.

healthy cells. But on its own a series of letters stored on a data server doesn't tell the whole story.

Of course, genes are important: they encode the recipes and instructions to make all the stuff in our cells and bodies, which are influenced by genetic variations and mutations. But as the saying goes, it's not what you've got, it's what you do with it that counts. How cells and organisms look, behave, and respond to the world around them—their phenotype—is created through a tangled web of genetic, epigenetic, and environmental interactions, switching genes on and off at the right (or wrong) time and in the right (or wrong) place. Then there's what I like to call the "wobble." Biology isn't a precise engineering project, and there's plenty of opportunity for random glitches and blips in the packed, bustling environment inside a cell, which can have an impact on the end result.

Frustratingly for people like Mel Greaves and the other researchers banging the drum for an evolutionary view of cancer, the past few decades of research have focused almost entirely on sequencing more and more genes and genomes. It's easy to measure genotype nowadays, all the way down to the level of a single cell. Measuring phenotype—how these genes are used and what they're doing—is a lot harder, although that's starting to change. One approach is mapping the patterns of gene activity in single cells by reading all the molecular messages that are produced when genes are switched on. We're getting closer to being able to do the same thing for the hundreds of thousands of proteins that are present in every cell.

Studying an individual cell's behavior and responses is even more difficult, especially as so much of this is dependent on the environmental context around it. A single cell plopped into a Petri dish is unlikely to behave in exactly the same way it would within the cellular society of a living organ, but researchers are only just starting to develop the tools that are needed to map this microscopic habitat in any kind of detail (page 200).

Importantly, evolution acts on phenotype, not genotype. Natural selection doesn't peer at a specific sequence of DNA inside a cell and say, "Yep, that gene looks like a good one, let's go with that." It acts on the output of those genes—how a cell or organism behaves and survives in its environment (its fitness), which determines whether it lives or dies and whether or not it gets to pass on those genes to the next generation. For all the talk of a genetic revolution wrought by high-speed, low-cost DNA sequencing, we're still struggling to open the black box between genotype and phenotype. It's proving to be hard enough to figure out how the millions of variations across the human genome interact with the environment in normal cells. Within the mutational mess of cancer, that task becomes many times harder.

While we might not be able to link all the molecular steps from genotype to phenotype in tumor cells, we do know the direction in which they're heading thanks to the Wheel of Death. More correctly known as the "Hallmarks of Cancer," this is a collection of the ten defining characteristics of the disease, first laid out in a neat circular diagram by American biologists Bob Weinberg (see chapter 4) and Douglas Hanahan. The Wheel of Death acts as a handy field guide to the phenotype of a successful tumor: self-sustained proliferation; ignoring stop signals; cheating death; renewing telomeres; disrupted metabolism; evading the immune system; chromosomal instability; invading and spreading around the body; growing a blood supply; and triggering inflammation.

Large-scale DNA sequencing has revealed a bewildering array of combinations and permutations of genetic changes that turn up in different kinds of cancers and healthy tissues at various points in time. But thinking of cancer as a collection of species evolving within the landscape of the body, ultimately heading toward the Wheel of Death, makes a lot more sense than seeing it as a shopping list of mutations and targets. Mutations are important, yes, but exactly which mutations matters less than the phenotype that they produce.

There could be many different ways to inactivate apoptosis or speed through the cell cycle in terms of the possible combinations of genetic and epigenetic alterations that might make it possible. From an evolutionary perspective, it's irrelevant *how* it happens—all that matters is that it *does*. And although it seems impossible to disentangle all the genetic complexity in cancer, knowing that the characteristics of individual cancers tend to converge on these hallmarks is a powerful insight into the underlying processes that may be at work. And if we can figure those out, we may be able to work out how to beat cancer at its own evolutionary game.

Imagine we could rewind back to the very earliest days of life on Earth, then play the tape again. Would multicellular organisms have still emerged? Would fish have still hauled themselves out of the sea to start a new life on land? Even if the dinosaurs had still mostly died out, would birds have evolved from what was left? Would *Hyracotherium* have become a horse? And would someone like me be here 4 billion years later, tapping these words into my computer?

It's the kind of scenario that often comes up in historical and cultural contexts, too. If someone went back in time and killed Hitler, where would we be today? Perhaps the horrors of World War II might never have happened, but the emergence of authoritarian fascist dictators does seem to be a remarkably common occurrence in any society throughout history that experiences a confluence of inequality, out-groups, polarized opinion, and a desire for strong leadership. If Hitler had been eliminated in the 1930s, maybe someone else with the same ideas and actions would have emerged from the maelstrom instead.

The question of what would happen if we ran the tape of life again is one of the longest-standing mysteries in evolutionary biology, occupying a philosophical as much as a scientific space.* It's impossible to follow the fate of a LUCA-like bug and its buddies in a rebooted primordial soup and recreate all the ecological tweaks, nudges, and flat-out catastrophes that happened over the 4-billion-year journey to today (not least because most funding bodies only tend to give out five-year grants). But cancer enables us to see the outcome of that evolutionary thought experiment again and again, on a timescale of single digits.

* This question is explored by Jonathan Losos in his fascinating book *Improbable Destinies* (Allen Lane, 2017).

Cancer starts from a human cell in a human body. There may be a lot of genetic permutations, environmental perturbations and selection pressures that it encounters en route to malignancy, but they're not infinite. Unlike the huge diversity of ecological niches on the planet—from sulphurous deep-sea vents to arid deserts—the habitats of the human body cover a more limited range. There will be variations between different organs, depending on their developmental history, but mammalian cells operate within a fairly limited range of conditions.

Looking at the natural world, we can see the same features evolving time and again as animals solve problems in the same way. Flying lemurs, flying squirrels, and sugar gliders all swoop between treetops thanks to a fleshy membrane called the patagium, which evolved separately in all three species. Animals tend to become unusually small if they're isolated on islands, regardless of where in the world they are, while plants become huge. Fossilized remains of unrelated species of miniaturized elephants have been found on islands in locations as diverse as the Mediterranean, the Bering Straits of the Pacific, the Californian coast, and the Indonesian archipelago in South East Asia. Porcupines on both sides of the Atlantic independently came to the evolutionary conclusion that the best way to avoid being eaten by predators was to grow massive spines. And Rong Li's super-yeast could only evolve in three different ways to solve the problem of life without Myosin II, even though each strain got there through a different combination of genetic changes (page 188).

It therefore makes sense that if two cellular environments

are similar, with the same genetic starting materials, there may only be a limited number of ways to reach the Wheel of Death. This might be achieved through mutations in specific genes, explaining why the same usual suspects seem to turn up so often in certain tumor types. This convergence even extends across species: the same handful of driver mutations turn up in a rare type of skin cancer in humans, dogs, and horses, suggesting there are very few evolutionary routes that this particular type of tumor can take.

In other cases, seemingly disparate cancers could be driven by disruption of the same biological network, built from the products of tens or even hundreds of genes. Combinations of alterations in any of these genes, whether strong or subtle, genetic or epigenetic, might be enough to send the system out of whack, without necessarily revealing an obvious driver mutation. Understanding and righting the underlying system imbalance is more likely to be effective at controlling the disease than trying to target a specific component of what might be a large and highly redundant system.

Although it seems like a formidable challenge, it's actually more doable than you might imagine. The human genome is big—around twenty thousand genes, plus more than a million control switches that turn them on and off—but there are only so many evolutionary escape routes and our genetic roadmap is growing more detailed every day.

Still, while genes and mutations are important—that's the fuel by which cancers evolve—every organism exists within and responds to its environment, and cancer cells are no exception. So maybe we can gain better insights into understanding, preventing, and treating cancer if we transport ourselves into the mindset of ecologists and evolutionary biologists, thinking about tumors as populations of genetically diverse individuals roaming around in the habitats of the body, subject to the rules and whims of natural selection. And in order to understand the evolutionary journey that each of them took to get there and figure out where they might be heading in the future, not only do we need to know about their genes, we also need to map the landscapes in which they live.

# 7

# EXPLORING PLANET CANCER

The view from up here is spectacular. Valleys and hills bursting with life, festooned with meandering rivers and bustling with a menagerie of curious creatures. Zoom out and it feels like you're flying over a verdant rainforest. Look closely and you might be able to spot a small group of predators gathering together before they head off to hunt. This is the landscape of cancer.

My guide is Yinyin Yuan, a talented young Chinese computer scientist and group leader within the Center for Evolution and Cancer at the ICR in London. She's using sophisticated image analysis algorithms to spy on the world inside a tumor, mapping out the ecological backdrop and the cellular species within it in exquisite detail. Yuan started her scientific life as a bioinformatician, sifting through reams of DNA sequencing data from cancer samples in the hope of understanding the genetic drivers of the disease. But she became disillusioned with the lack of detail about the organization and architecture within tumors. She didn't just want to know about the shopping list of mutations that might be present in a chunk of tissue containing millions of cancer cells—she wanted to know where these cells were, how they were arranged, and what else might be in there, too.

In the rush to sequence as many cancer genomes as possible, we've forgotten that the tissues of the human body and the tumors that grow within it are a multitude of miniature habitats, each with its own characteristics. Viewed in this way, it's obvious that genetically mixed-up cancer cells will adapt and evolve in different ways according to the environment in which they're growing, just like species in the natural world.

Studying cancer genomes in a test tube of mashed-up tumor chunks—which contain healthy cells, immune cells, cancer cells, and more, all inhabiting their own ecological niches—reduces all this diversity to a flat beige territory bereft of any distinctive features. An ecologist wouldn't go to the bottom of a valley and look at all the lush greenery growing alongside the river, climb two miles to the very top of a mountain to spy on tiny alpine heathers, and then draw a map of a flat plain with a height of half a mile where weeping willows and edelweiss grow side by side. It would be scientific nonsense, yet this is exactly how we've been studying cancer in recent years.

Pathologists have been peering down their microscopes at wafer-thin slices of tumors for more than a century. Even today, the standard way of diagnosing a cancer is to pop it under the lens and look at the cells. But unlike the race for ever faster and cheaper DNA sequencing of more and more cancer genomes, pathology has been left lagging. It's only in the last few years that technology has advanced to the point where it's possible to turn these images into the kind of detailed digitized data that Yuan wants to explore, analyzing the cells and conditions *in situ* within a tumor.

She pulls up a high-resolution microscopy image onto her screen. It's a slice through a lung tumor, collected as part of Charles Swanton's TRACERx study (page 173). The multicolored landscape looks for all the world like a satellite image of a rainforest, punched through with two empty holes where samples have been taken for DNA sequencing and other types of molecular analysis. I watch in fascination as she zooms in and out as if we're exploring the world with Google Earth. We pause for a moment on a blood vessel snaking its way through a clump of cancer cells, then hover over a cluster of immune cells lurking on the edge of a tumor in astonishing detail—a whole panorama captured within a wispy slice of tumor roughly the diameter of a penny.

I can't help thinking how far we've come since the days of the nineteenth-century microscopists and their neat pencil sketches of malignant cells. It would have blown Theodor Boveri's mind to know that today's pathologists can put 300 glass slides plastered with thin slices of tissue into a digital scanner overnight and come back to a hard drive full of high-resolution images in the morning. And instead of painstakingly cataloging and counting cells by hand, computer algorithms are being trained to do the hard work of identifying and mapping all the features within these biological landscapes.

Yuan and her team are attempting to address the fundamental problem with our current approach to characterizing cancer. Simply mashing up and sequencing a tumor sample instantly erases all the information about the spatial arrangement of cancer cells and their normal neighbors. Instead, she's using her advanced imaging techniques and AI algorithms like flying a drone over dense jungle, capturing the organization and diversity of all the species within it.

It's not just the details of the genetic patchwork that get missed with the current smash-and-sequence approach. Researchers are increasingly paying close attention to the distribution of immune cells in cancers, whether as lethal predators or inflammatory agents provocateurs. But while there are flashy cell-sorting machines that can tell you how many immune cells are present in a mashed-up tumor sample, they reveal nothing about where they are and what they might be up to.

It's been known for a while that the greater the number of immune cells within a tumor, the more likely it is to respond to treatment. But, as Yuan and her team are now discovering, what really matters is where they are. By way of example, imagine you're looking at a series of five aerial reconnaissance photos from a forest, trying to figure out the distribution of pesky invasive mice and predatory hawks. In four of the pictures there are plenty of birds hunting and hovering, keeping the pests in check. Based on an average of these images it might look like the War on Mice is being won. But the final picture tells a very different story: it's a pocket of dense woodland that makes for tricky hunting territory. This quickly becomes overrun by little squeakers that will doubtlessly be spreading into neighboring areas in search of new homes.

Now imagine these pictures are high-resolution images of the landscapes in five samples taken from different parts of a tumor, with immune cells as the hawks and cancer cells as the mice. Say you have one that contains 10 percent immune cells, two with 20 percent, another at 40 percent, and one with just 5 percent. These figures would usually be averaged out to give a figure of roughly 20 percent immune infiltration (as it's technically called). But that ignores the fact that there's one "hot" region that's boiling over with immune cells and one that's barely bothered by their presence. This "cold" zone effectively creates a predator-free haven for cancer cells to proliferate unchecked and potentially start migrating around the body to start secondary tumors elsewhere.

Yuan and her team have now taken this idea a step further, developing an algorithm that analyzes the number of hot and cold spots across multiple samples from patients with lung cancer, finding that tumors with a larger number of immune cold spots are more likely to come back after treatment. It's something that would never have been obvious from just averaging the number of immune cells across a whole tumor, but it could be vital information for doctors trying to work out the best strategy for treatment.

Importantly, the immune system has the capacity to adapt and evolve in the same way that cancer does, holding potential for long-term control or even cure. So it's no surprise that immunotherapy is probably the hottest topic in cancer treatment right now. Front of the pack are drugs known as checkpoint inhibitors, which alert immune cells to the presence of cancer and trigger them to attack. They've been particularly successful in melanoma and lung cancers, with some patients even being apparently cured of terminal disease, although they only work for fewer than two in five of the patients who get them. Some of this disparity in response can be blamed on the non-human cells living inside our bodies—the billions of bacteria, fungi, and viruses collectively known as the microbiome—and people whose cancer responds to immunotherapy appear to have a different selection of gut microbes compared with those who don't. There's a lot of work going into finding ways of boosting the effectiveness of immunotherapy and devising tests to work out who will benefit the most, although some oncologists will offer immunotherapy as a last-ditch option even without that information.

The idea of growing genetically modified immune cells in the lab is also starting to gain more traction. Researchers are using new gene editing techniques such as CRISPR to create universal "super-soldiers" that are trained to seek and destroy cancer cells, although these treatments are extremely expensive owing to the technology involved. A couple of modified immune cell products (known as CAR-T cells) are on the market for blood cancers, with plenty more exciting innovations coming hot on their heels, but these treatments are still in the early stages of testing for solid tumors at the time of writing.

However, the immune system is a powerful beast and great care should be taken when waking it. We don't know anything about the long-term impacts of activating the immune system in this new way, either good or bad. Overstimulation can cause a so-called cytokine storm that floods the body with immune signals, triggering a massive reaction that results in severe side effects or even death. Overenthusiastic immune cells can also turn on the healthy tissues of the body, attacking the nerves, gut, and skin. A number of stories have also started to emerge about the phenomenon of hyper-progression, where immunotherapy triggers an explosion of rapid tumor growth. All of this has to be balanced against the limited chance of success, to avoid bringing unnecessary suffering to someone who's fast running out of time and options.

There are many different types of immune cells with a whole range of jobs, adding another layer of complexity. Some are predatory hunters while others are inflammatory troublemakers. There are peacemakers, which damp down excessively aggressive immune attacks, and there are garbage disposal units that gobble up the cellular casualties. Some of these responses are useful for controlling cancer, but others can encourage tumor growth. At the same time, highly predatory immune cells can be a potent selective force shaping the evolutionary direction of a tumor. Results from the TRACERx team show that some lung cancers go into "stealth mode" in response to sustained immune attacks early on in tumor development, switching off or losing the distinctive molecular flags that would otherwise attract the attention of predatory cells.

Though there may be many types of immune cells in and

around tumors, there are vastly more cancer cells, each carrying its own assortment of genetic mutations. Yinyin Yuan's next big challenge is to find out more about the species of cancer cells that thrive in the different habitats within a tumor and map that onto what we know about their genetic, physical, and behavioral traits. Connecting the wealth of DNA data coming from the sequencing labs to the underlying ecology of the disease is a highly complex task, but it's one that can now be addressed thanks to advances in artificial intelligence and machine learning.

It makes sense that the diversity that she sees down the microscope should be reflected in the underlying genetic diversity, but matching cancer cells to their habitats is a complicated job. It'll require high-resolution spatial analysis of DNA, proteins, cell characteristics, and patterns of gene activity, along with computational tools to crunch all the information together and make sense of it. Gathering all this data is going to be much harder than simply shoving samples of mashed-up tumor through a sequencing machine. It's also going to need a huge amount of computing power and storage space, given how much larger image files are than reams of DNA letters. But it's also going to be much more informative for understanding the complex ecology of cancer and tackling it more effectively.

Yuan and her team are busy working on ways to turn their algorithms into clinical tools for doctors, mapping out the molecular landscape of a tumor and the cellular species within it in order to help doctors pick the best options for treatment. There are some other fascinating technical innovations on the way, including the ability to analyze patterns of genetic activity in single cancer cells within a slice of tumor and maybe even one day sequence their genomes, too. Automated image analysis techniques, known as computer vision, also have the potential to revolutionize the field. For example, researchers at New York University recently hacked one of Google's off-the-shelf image analysis algorithms to identify the likely mutations in clusters of lung cancer cells purely from looking at microscopy images. It's not 100 percent accurate, but it's a promising start.

One more point that's worth mentioning: the cancer samples that are used for DNA sequencing and image analysis are just a snapshot of dead biology. They're a tiny slice from the genetic fossil record of a tumor that's spent years dynamically evolving and adapting in response to changes in its environment. So trying to reconstruct the molecular complexity of a tumor from a couple of samples is like trying to infer every single genetic change on the journey from *Hyracotherium* to horse just by looking at skeletons of two ancient and modern specimens in a museum.

Catching cancer evolution in action by studying samples taken at every point from initiation to fatal metastatic disease would enable us to understand what happens on the journey from sad cell to bad cell as a tumor starts. It would also allow us to follow the journey of pioneering cells heading off into the bloodstream and establishing secondary cancers, and see what happens as cancers shrink and bounce back in response to therapy.

It's not space that's the final frontier in cancer research—it's time.

## THE WOUND THAT DOES NOT HEAL

So far we've focused on cancer cells as passive players in their environment, adapting and responding to changes in the world around them. But this is only half the story. Not only does the environment shape species; species shape their environment, too.

All over the world we can see the impact that organisms have made on their surroundings, from the industrious dam-building of beavers or the rampant tunneling of rabbits to the unmistakable ecological boot prints left by humankind. Much in the same way that we humans have altered our environment to support our survival by building shelters, harnessing fire, and farming, cancer cells are dystopian engineers, destroying the orderly neighborhood in which normal cells thrive and constructing a *Mad Max*-style setting that's better suited to their greedy, cheating nature.

Although Yinyin Yuan's images of tumors may look like flourishing rainforests, the reality is much less pleasant. Cancer cells are like the very worst neighbors you can imagine, sucking away all the oxygen and nutrients, and polluting the environment with their waste. Healthy cells generate energy by "burning" oxygen and glucose (a type of sugar) through a complex chain of biochemical reactions known as aerobic respiration, producing water and carbon dioxide as by-products. Fast-growing tumors are low-oxygen environments thanks to their chaotic blood supply (page 216), so cancer cells often switch to an alternative process known as glycolysis—an oxygen-free metabolic pathway that originally evolved in ancient bacteria living in suffocating deep seas. Cancer cells using glycolysis burn through glucose ten times faster than healthy cells and throw out waste in the form of lactic acid, quickly turning healthy tissue into a noxious wasteland.

The switch to glycolysis is more commonly known as the Warburg effect, named after the German biologist Otto Warburg who first noticed that cancers preferred to burn sugar in the absence of oxygen in the 1920s. He passionately believed it was the metabolic change and acidification that caused cancer, rather than the other way round. Warburg won a Nobel Prize in 1931 for his work on energy production, and was highly dismissive of what he considered to be wasted effort on studying carcinogens, genes, and cancer viruses. His ideas were roundly rejected once it was discovered that tumors are driven by the accumulation of genetic faults, as there was no obvious way in which altered metabolism could cause mutations.

Despite the rampant proliferation of conspiratorial websites and YouTube channels lionizing Warburg as the founder of the "Truth" about cancer that "They" don't want you to know, there's still no solid evidence to show that altered metabolism leads to malignancy. Harking back to the idea of adaptive oncogenesis (page 122), where wannabe cancer cells will thrive if they're better suited to stressful surroundings than their healthy neighbors, it's easy to see that a highly acidic, low-oxygen environment could act as a strong evolutionary pressure. Flipping to Warburg-style glycolysis and acidification might only happen in a small pocket of cancer cells—perhaps as a result of a temporary oxygen shortage—but it might be enough to support the emergence of hardy mutants that can survive in such harsh conditions.

As we explore this toxic environment, it's important to remember that there's more to a tumor than cancer cells. There's a whole bunch of other cells and stuff, collectively referred to as the stroma. This is a diverse selection of immune cells of all flavors, "filler" fibroblast cells, and blood vessels, all stuck together with molecular glue (the extracellular matrix) and soaked in a bath of biochemical signals. A typical pancreatic tumor might be made up of as little as 10 percent cancer cells, with the rest comprising normal cells that have been co-opted into either supporting or fighting against them. But although this strange community is clearly vital for cancer cells' survival, little is known about how they exert such a corrupting effect on their surroundings.

The secret may lie in the ability of cancer cells to hack the normal biological processes of the body and re-engineer them to suit their own needs. For example, healthy tissue reacts to the presence of a tumor as it would to any injury, sending in the immune troops to fight off the invaders, triggering inflammation to promote healing and knitting the tissue back together with stringy fibroblasts. But it's becoming increasingly clear that tumors are manipulating this good-natured healing process in order to fuel their growth.

One clue as to how they're doing it comes from biologist Gerard Evan and his team at the University of Cambridge, who have spent decades investigating a key cancer driver gene called *MYC*, which is overactive in many different tumor types. Under normal circumstances, *MYC* helps to kickstart the complex biological ballet required for wound healing and tissue regeneration, and it's switched off once its job is done. But activating *MYC* non-stop in precancerous lung cells is enough to spark a corrupted recapitulation of the normal healing process, driving benign "sad" cells to multiply into an aggressive "bad" cancer.

Given that inflammation can disrupt the orderly society of cells, it makes sense that anything that triggers or increases the intensity of these processes—such as tissue damage or inflammation—would also fuel tumor growth. Conversely, finding ways to control inflammation and make wound healing run to neat completion rather than continuing to fester chaotically could be a useful approach for treating cancer.

This idea of tumors tapping into the normal processes of life and corrupting them for their own needs also extends to hacking the biological plumbing of the blood system. Since the earliest days of medicine, doctors have noticed that tumors are fed by their own blood supply. The common interpretation of the origin of the very name "cancer" comes from Hippocrates' use of the Greek word for crab—*karkinos*—back in the fifth century BCE, supposedly in reference to the bulging veins sprawling out of a tumor like a crab's legs, or maybe as a metaphor for the disease's pincer-like grip on the body.*

* There's an alternative etymology first pointed out by Italian-English doctor Louis Sambon in the 1920s, based on the behavior of a parasite known as *Sacculina*. Normally a free-swimming barnacle-like creature, *Sacculina* also makes a living by burrowing into the bellies of passing crabs, creating tumor-like masses. Noting the ancient Greeks' fondness for crustaceans, whether on their plates or depicted in art and jewelry, Sambon couldn't believe that Hippocrates would have missed the obvi-

The growth of new blood vessels (a process known as angiogenesis) is a vital step in the journey from a blob of selfish cells to a full-blown tumor: oxygen and nutrients can only permeate through a few hundred cells on their own, so a cancer can't grow much bigger than the period at the end of this sentence without growing a blood supply. And, of course, blood vessels provide a very handy route for metastasis, allowing cancer cells to escape the tumor in which they were born and head off to distant parts of the body.

Unlike the orderly pipework running through healthy tissue, the blood vessels in cancer are a disorganized, chaotic mess, snaking their way into tumors in response to molecular distress calls from starving, suffocating cells. In 1971, an ambitious young Boston-based surgeon named Judah Folkman discovered that cancer cells produced a soluble chemical that induced the growth of new blood vessels when injected under the skin of rats, which he called Tumor Angiogenesis Factor. Enthused by his results, he tried to persuade the research community to start developing drugs that interfered with this mysterious Factor in the hope of shutting off the blood supply to thirsty tumors and stopping them from growing to a dangerous size.

His ideas were initially met with skepticism by many of his peers, who believed that tumors grew alongside existing blood vessels rather than plumbing in their own supply. More than a decade later, Folkman was vindicated when

---

ous analogy between these strange lumps and the deadly tumors affecting his human patients.

researchers unveiled the identities of a number of molecules that are produced by cancer cells and stimulate new tubes to sprout from nearby blood vessels.

One of the most exciting of these was Vascular Endothelial Growth Factor (VEGF), which quickly became a target for researchers eager to exploit ways of shutting off cancer's blood supply. Front of the pack was Californian biotech company Genentech, who spent most of the 1990s developing the first VEGF-blocking drug, Avastin (bevacizumab). There was plenty of hope that this would be the breakthrough the world had been waiting for, with no less an authority than James Watson (co-discoverer of the structure of DNA) declaring that Folkman's discovery was "going to cure cancer in two years."

Folkman himself was much more cautious, pointing out that the drugs had only been tested in animals and were yet to go into human trials.* Despite the hype, Avastin has largely been a bust as far as treating cancer is concerned, although it has had a second lease of life as a treatment for macular degeneration, a type of progressive sight loss driven by new blood vessels growing in the back of the eye.

The explanation for Avastin's failure to live up to these lofty expectations dates back to 1999, five years before the drug was first approved for treating cancer. Deep in the Anatomy Department at the University of Iowa, cell biologist Mary Hendrix and her colleagues were using high-powered microscopy to take a closer look at the structures inside

* He also went on to say, "If you have cancer and you are a mouse, we can take good care of you"—a statement that's as true today as it ever was.

melanomas, focusing in on the meandering blood-filled loops making their way through the mass of cancer cells. Contrary to the angiogenesis dogma of the time, which held that this disorderly pipework must have sprouted from nearby blood vessels, the capillaries inside these melanoma samples were actually built out of repurposed tumor cells—a phenomenon Hendrix described as vascular mimicry. The discovery that cancer cells could morph into blood vessels and plumb themselves back into the bloodstream rather than sending out signals to encourage new vessels to come hither was as controversial as it was startling. Few people believed the heresy, even as one anti-angiogenesis drug after another flopped in clinical trials.

Hendrix and a few other believers kept the faith, publishing a handful of papers every year showing that vascular mimicry was real, important, and worth thinking about. It wasn't until 2015, when molecular biologist Greg Hannon at the Cancer Research UK Cambridge Institute published a paper in the journal *Nature* showing the same thing that the idea started to gain more widespread acceptance. Hannon and his team found that breast cancer cells transplanted into mice could convert themselves into blood vessels, plumbing the tumor back into the mains and providing a conduit for the disease to spread.

The discovery that tumors can tap into underlying biological programs and plumb themselves a new blood supply isn't entirely unexpected. All the same genes that are used to build a blood system in a developing embryo are still present in adult cells, although they're switched off, so why not use them? More importantly, the discovery that cells can organize into specific roles within a tumor tells us that it isn't a completely anarchic mess of selfish cells. It's a horrible place that favors rebels and cheats over well-behaved healthy cells, but there's still a level of organization within this dystopian society.

Researchers have even found examples of cooperation between clusters of cancer cells in a tumor, each producing molecules that enable their neighbors to survive. It might seem like strange behavior from cellular scoundrels, but it's less odd when viewed through an evolutionary filter. Multicellularity has emerged multiple times in history as selfish single cells teamed up and specialized into their own roles within the collective body. So we shouldn't be surprised to see the same journey repeated in the microcosm of cancer.

## A TALE OF MONKEYS AND METASTASIS

A long, long time ago a very, very small number of very, very lucky monkeys made an impossible voyage. The exact circumstances are lost in the mists of ancient history, so we'll have to imagine the heroine of our story is a pregnant female, playing about on a mat of leaves and twigs in the mouth of a river on the African coastline only to find herself swept out into the fierce Atlantic by an unexpected storm. Pushed by a fast tide and a following wind, her makeshift raft washes up on a foreign coast. She's a long way from home, half-dead from exhaustion and starvation, but she's alive. Judging by the kicks in her belly, so are her twins. Fast-forward 36 million years or so and every single one of the 100 species of native monkeys in the Americas is her descendant.

This story sounds wildly implausible, but genetic analysis of the relationships between all the New World monkeys tells us that at least some version of it must be true. Maybe the founder was our plucky pregnant mom of two, cast adrift on her vegetable raft. Maybe a small family hopped their way across the Atlantic along a chain of now vanished islands. Whatever happened, it only happened once. But once was enough.

Here's another implausible story. A single gram of invasive tumor contains up to a billion cancer cells, many of which are continually being shed into the bloodstream. At this point, it may seem that secondary cancers are a grim inevitability for every patient, but the odds are much smaller than you might think. A teaspoonful of blood from someone with cancer typically contains fewer than fifty tumor cells, meaning that there are a couple of tens of thousands swilling around in the full 1.5 gallons of red stuff at any time.

Most of these cells only make one trip around the body before they're destroyed, but it still adds up to several million cells a day or a few billion every year. This figure is extraordinarily high considering the handful of secondary tumors that turn up in patients, telling us that the ability of any single cancer cell to seed a metastasis is one in a billion or thereabouts. The process of a cancer spreading throughout the body is mathematically implausible. Yet it's a biological reality and, given the numbers involved, even the most improbable odds will lead to a "winner" at some point.

If tumors stayed in one place, we would be able to cure

nearly all solid cancers with nothing more than sharp surgical steel. The single thing that makes the disease so lethal is metastasis—cells breaking away from a primary cancer, traversing the highways and byways of the body, and seeding secondary tumors in new locations. First comes invasion, when tumor cells break through the membrane barriers that line tissues and organs. It's a short hop from there to freedom, either tumbling through blood vessels or finding a path through the lymphatic system—the network of ducts and nodes that form a secret highway for immune cells. If a tumor is removed before the first invasive breakthrough, then a cure is almost certain. Unfortunately, once that barrier has been breached then it's pretty much guaranteed that cancer cells will have started to circulate by the time a tumor is large enough to be detected using conventional diagnostic techniques.

By the tail end of the nineteenth century, anesthetic and antiseptic techniques had improved to the point where surgery for cancer was feasible, if still very risky. Breast tumors were particularly amenable, being located in an accessible and dispensable part of the body rather than buried deep within a vital organ. Frustratingly, while some women with breast cancer were indeed cured by the knife, many others succumbed to secondary tumors that sprang up in the bones, lungs, liver, and brain. Concluding that this must be the work of wandering cancer cells that had escaped from the original tumor, pioneering American surgeon William Halsted invented the radical mastectomy—a new operation that involved removing the breast along with the underlying muscles and lymph nodes in the armpit.

Halsted's strict adherence to antiseptic technique undoubtedly saved lives from infection and his dedication to pain relief made the grueling operations bearable (accidentally leading to a personal addiction to cocaine and morphine along the way). But his extensive surgery didn't make much of a difference to survival from cancer. Some women were cured, others succumbed to secondary tumors.

The solution of the surgeons that followed in his footsteps through the early twentieth century was to go even further. Their "super-radical" mastectomies were little short of brutal, cutting away more and more flesh in an attempt to stem the spread. Ever deeper layers of muscles were stripped away and some women ended up having parts of their shoulder or arm amputated. Yet, despite the deeply distressing physical and psychological damage inflicted on women through these operations, which continued through to the 1950s, survival rates remained unchanged.

This wave of mutilation became too much to bear. A small group of surgeons in Europe and the United States fought back, gathering enough data to prove beyond doubt that the chances of developing secondary cancer were the same regardless of whether women underwent a mutilating radical mastectomy or a smaller operation to remove just the affected breast. The message was clear: by the time the cancer was obvious enough for an operation, there was a risk that the seeds had already been sown. Some women would be lucky, others wouldn't. But hacking at their bodies wasn't the solution.

Around the same time that Halstead was slicing Stateside

in the hope of stopping metastasis, British surgeon Stephen Paget was wondering why it happened at all. Poring over the autopsy notes from more than 700 women who had died of metastatic breast cancer, he noticed that the disease tended to spread to some organs and not others. Why should cancer cells prefer to set up their second homes in the bones and lungs, and not in the spleen? And if the appearance of secondary cancers is the result of clumps of cancer cells getting stuck in tiny blood vessels, as was commonly thought at the time, then why should the liver be a target and not the kidney, which does the dirty work of filtering gallons of blood every single day? And why should other types of cancer have different preferences? Paget explained his observations through the analogy of "seed" and "soil," writing "When a plant goes to seed, its seeds are carried in all directions; but they can only live and grow if they fall on congenial soil."

Over the past century we've come to understand the molecular nature of Paget's metaphorical seeds in ever greater detail, mapping the mutations in cancer cells, and dissecting the genetic differences between tumors that spread and those that don't. Tracing the family trees of primary tumors and distant metastases has revealed some secondary cancers are seeded from just one or a handful of genetically identical cells, like the adventurous monkeys that made it to South America, while others grow from a more diverse group of cells moving in together. These wandering cells don't necessarily have to have come from a large primary tumor, either. One study looking at more than a hundred tumor samples collected from twenty-three patients with bowel cancer that had spread to their liver or brain showed that in four out of five cases the seeds of metastasis were probably first sown when their original bowel tumor was no bigger than the head of a pin.

Occasionally, cancer cells relocate for a second time, breaking away from secondary tumors and growing in new locations. Sometimes they'll even leave a primary tumor, take a fantastic voyage around the body, and come back home again. This rehoming ability could potentially be exploited for treatment—researchers have been working on ways to capture wandering cancer cells in the bloodstream and use genetic engineering tools to turn them into lethal double agents, primed to kill their comrades when they return to base.

Far less progress has been made in figuring out what determines the congeniality of the soil in which these seeds fall. The question of why some cancers prefer to spread into the bones while others feel more at home in the brain, liver, or lungs remains one of the greatest mysteries in the field, although some of these secrets are finally starting to be unearthed.

Not only do cancer cells corrupt the healthy tissue immediately around them, but they also brainwash normal blood stem cells to act as their ambassadors. These manipulated cells cluster together in nooks and crannies within the bones and organs of the body, receiving instructions in the form of chemicals released into the bloodstream from their primary tumor commander and preparing a comfortable new home for any migrating cancer cells that should be passing by. However, there's still no guarantee that these immigrants will prosper in their new locale.

It's now known that there are many tiny micrometastases lurking in the bodies of most cancer patients, although the vast majority of them will never develop into a secondary tumor. The same principles of the cellular society are still at work, with well-ordered healthy tissue keeping cheating cancer cells under control wherever they end up. It therefore also makes sense that inflamed, damaged, or aging tissue might be an attractive location for wandering cancer cells or encourage tiny dormant tumors to awaken and start growing.

An intriguing insight into how this might work came at the end of 2018 from a team of scientists based at Cold Spring Harbor Laboratory in New York. Using mice as a model for human tumors, they scattered single breast cancer cells into the animals' lungs (a frequent location for metastasis) and waited to see whether or not any of them grew into secondary tumors. After eight months they'd got nothing. Nada. The cancer cells were still there, but they were lying dormant in a state of suspended animation.

Next, the researchers dosed the animals with bacterial chemicals, mimicking the effects of a nasty lung infection. Immediately, the cancer cells started growing into new tumors. Exactly the same thing happened when they exposed the mice to cigarette smoke. Curiously, neither the bug juice nor the smoke had a direct effect on the cancer cells. Instead, these insults activated special inflammatory immune cells called neutrophils, causing them to throw out microscopic molecular nets of DNA and proteins. Caught in the neutrophils' tangled web, cancer cells wake up and fight back by proliferating as fast as they can.

Intriguingly, a drug that blocks one of the molecular signals that triggers inflammation has already been tested in a large-scale trial for cardiovascular disease, in the hope of calming the inflamed, clogged arteries that cause heart attacks and strokes. Not only did the drug reduce the chances of dying from heart disease, but the people receiving the drug also had a much lower incidence of lung cancer than might have been expected, even though many of them were smokers. However, a different study revealed that common steroid anti-inflammatory drugs called glucocorticoids might exacerbate the spread of breast cancer. Despite the growing interest in anti-inflammatory drugs, these conflicting results suggest there's still a lot we don't understand about what turns wandering cancer cells into metastatic tumors and how to stop them.

It's also interesting to consider why cancer cells might start going on the move in the first place. Maybe it's purely accidental—loose cells on the edge of a tumor washed away on the passing tide in the bloodstream. But many of the millions of cells that migrate away from a tumor every single day are also actively trying to leave.

From the point of view of an ecologist, the emergence of metastasis in a cancer makes perfect sense. Animals—including humans—migrate in search of resources such as food or space when things get difficult on their home turf, risking their lives over thousands of miles in the hope of making a better life for them or their offspring. Species can also be forced away by an intolerable climate or unsafe conditions. In the tightly packed, toxic tumor environment only the very best-adapted, hardiest cells will survive. Many will die, but as food and oxygen start to run low and immune predators move in, some will choose to take their chances elsewhere.

Cells aren't sentient so this isn't an active "choice" in the same way that we would understand it, but is driven by fundamental biological motivations. For example, cells can sense the levels of various chemicals in their environment—including sugar, oxygen, and molecular building blocks called amino acids—and will tend to move toward areas with more of these goodies if they're able. This also suggests that the well-intended but misguided idea of trying to "starve" cancer by manipulating the availability of nutrients such as sugar may actually do more harm than good. As well as being very difficult to achieve, because our bodies have evolved to be extremely good at extracting usable sugars from all kinds of sources, attempting to starve out cancer might actually make the disease more likely to spread as the cells go in search of nutrients. And instructing patients to cut out easily-digestible sugars could end up depriving them of a valuable energy source that they need for healing.

As a die-hard biologist, it pains me to admit that the physical sciences may also have some lessons to teach us about metastasis. As technology improves to the point where we can probe the physical properties of biological structures, scientists are starting to explore how this affects the ability of cancer to grow and spread.

Paradoxically, it turns out that cancer cells are actually softer and more malleable than healthy ones, even though tumors usually reveal themselves as hard lumps within squishier normal tissue. This is down to all the fibroblast fillers and the sticky extracellular matrix of the tumor stroma, providing a firm foundation through which the softer cancer cells can move more easily—imagine a runner sprinting along a damp, firm beach instead of trudging through dry, shifting sand dunes.

Although there's a lot of focus on the genetic errors and molecular signals that drive metastasis, the answer may be as simple as the shape of the cancer cells themselves. Imagine you're having a night out in a jam-packed pub and have finally managed to navigate your way to the bar. Clutching a pint in each hand, you now have to make it back to your drinking buddy on the other side of the room. Most people will instinctively turn sideways, making a narrower shape that slips more easily through the crowd. According to physicist Josef Käs from the University of Leipzig in Germany, cancer cells do exactly the same thing. By taking careful measurements of cancer cells growing within the sturdy framework of the stroma, he and his team have found they get elongated and squashed, enabling them to slip sideways through the crowded tumor environment.

When I saw him present his work at a meeting in London, Käs showed a video of chunky hexagonal plastic blocks being shaken from side to side in a tray, mimicking the tightly packed cells of the stroma. If all the blocks are the same shape, then the "cells" stay put, wobbling slightly with the movement of the tray. But stick a couple of longer blocks in there to represent squashed cancer cells and the whole lot starts to mobilize. Eventually, the elongated cells work their way free and pop out—they have "metastasized" using nothing more than their shape and the physical interactions that happen as they jiggle around. Perhaps, he argues, there could be ways to jam up this fluid movement, "freezing" cancer cells in place and preventing them from spreading around the body. It's a weird idea, but it might just work.

## TAKING A BROADER VIEW

Zooming out even further, we start to see that a tumor isn't a rarified, encapsulated environment like the isolated biomes of Cornwall's Eden Project but an integral part of the body. To misquote the poet John Donne, no tumor is an island, entire of itself. It is an inextricable part of the larger continent, locked in a two-way union that's impossible to unpick.

For starters, cancer cells are exposed to the contents of the hormonal soup within the body and may also churn out hormones themselves. There are sex hormones—principally estrogen and testosterone—which are implicated in many breast and prostate cancers respectively. Other hormonal players include insulin-like growth factors, which control how the body uses energy and stores fat, potentially providing an explanation for why being overweight brings an increased risk of some types of cancer.

One of the key players here is IGF-1, a growth factor related to insulin. The remote province of Loja in southern Ecuador is populated by little people, all around three feet in height. Although they may be short, they live unusually long lives and are resistant to cancer, diabetes, and many other diseases. Known as Larons, after Israeli doctor Zvi Laron who first described the syndrome, these people have a hereditary gene fault that knocks out their IGF-1 production. Understandably, there's a lot of interest in experimental low-calorie diets that create a similar drop in IGF-1, in the hope that the rest of us might benefit from the Larons' genetic secret.

The idea of dietary restriction for longevity was first put forward by sixteenth-century Venetian Alvise Cornaro, whose *Writings on the Sober Life: The Art and Grace of Living Long* recommends that would-be centenarians should eat just 350 g of food per day (around 1,000 calories), consisting of bread, soup with egg, meat, poultry, and fish, along with a little under half a liter of wine.

More recent scientific research suggests that Cornaro may have been on to something, despite his interesting interpretation of sobriety. Restricting calories does seem to affect healthspan—the length of time someone maintains good health and vitality—perhaps by keeping the body's internal habitat in good shape, even if the evidence on lifespan is mixed. However, underfeeding might also affect the micro-environment of the body if there aren't enough nutrients for proper tissue repair and maintenance. Severe diets can lead to restricted growth, flagging energy levels, a loss of libido, and may even dampen down potentially cancer-fighting immune responses. As the joke goes, dieting doesn't necessarily make you live longer—it just feels like it . . .

There's also a potential role for the billions of microbes living on and in the body, collectively known as the microbiome.

Previously thought to have little relevance to cancer research, the microbiome has recently emerged as one of the hottest of hot topics. Researchers have discovered that the bacteria living in the gut can affect how well someone's cancer will respond to chemo- or immunotherapy. Others have found that exposure to the right kinds of bugs in early life protects children from developing leukemia. Some types of bacteria contribute to the growth of cancer in places like the liver and bowel, while specific fungal infections have recently been shown to play a part in promoting the development of pancreatic cancer. More directly, microbes in the gut can alter the availability of certain nutrients (either for the healthy tissues of the body or any tumors), produce potentially carcinogenic chemicals, or even manipulate the immune response—all of which could affect the development, progression, and treatment of cancer.

The connection between the body clock and cancer is another underexplored area. Every part of your body runs on a daily cycle, kept in sync by a small cluster of nerve cells in the brain that act as a master clock. Not only does our body clock dictate when we feel alert, sleepy, or hungry, but it also dictates the best times of the day for our cells to renew and repair themselves. So it stands to reason that messing around with the body clock might have an impact on cancer risk.

In 2007, the International Agency for Research on Cancer (IARC) classed shift work as "probably carcinogenic to humans," although this continues to be a hotly debated topic as new studies emerge. There's also the intriguing idea of chronotherapy: delivering treatment according to the body clock by giving DNA-damaging drugs or radiotherapy at times of the day when cancer cells are less able to repair themselves.

The word "holistic" may have been relentlessly abused over the years by wellness gurus and alternative therapists, but I think it's time it was reclaimed by science. The study of cancer has become overly reductionist, focusing on genetics and genomics right down to the level of single cells. But knowing the exact genome of one mouse scuttling through a forest, which may survive or die within a season, tells you little about the behavior of the whole species and how it changes over time, or how the population interacts with all the other woodland creatures. It's time for a more holistic understanding of the cellular species involved in cancer, the ecology of its habitats and its evolutionary journey rather than the laser focus on specific mutations, drivers, and targets.

We also need to step back and view the disease as a complex, ever evolving ecosystem within the wider world of the body, capable of throwing up all kinds of diversity and innovation in an attempt to adapt and survive. And that includes some *really* weird stuff.

# 8

# SURVIVAL OF THE WEIRDEST

A mathematician by training, Kristin Swanson is an unconventional professor of neurosurgery at the Mayo Clinic in Phoenix, Arizona, wielding equations with the same skill and precision that her surgical colleagues wield their scalpels. Over the past fifteen years she's been building a database of nearly 3,000 patients with brain tumors, scraping every pixel of data from their MRI scans to build mathematical models that can help to predict how their cancers will grow and the best option for treatment.

While sifting through all this information, she noticed something strange: tumors from male patients tended to keep on growing through chemo- and radiotherapy, while those from women slowed down in response to treatment. It wasn't 100 percent true in every case, but the difference between the sexes was clear enough. Swanson became intrigued and started looking for an explanation.

There are some fundamental biological differences in

cancers that affect men and women. Most obviously, there are differences in anatomy that affect the chances of getting certain cancers. If you have a cervix you can get cervical cancer, if you don't then you can't. The same goes for ovaries, uteruses, testicles, and prostate glands.* There are also discrepancies in the incidence of cancers affecting both sexes, with men having a greater risk of the disease overall. Although some of that is down to lifestyle and habits (men tending to smoke and drink more, for example), it's not enough to explain the difference.

One reason might be hormones, which vary between the sexes and fluctuate throughout life. Another explanation could be the sex chromosomes themselves. Genetically speaking, people who have two X chromosomes are female while those with an X and a Y are male. The Y has a tenth as many genes as the X, is around a third of the size of the X and has a habit of accidentally go missing when cells divide. This is known to happen in blood cells of older men, especially those who smoke, and seems to be linked to a higher risk of various types of cancer.

Statistics show that men are more likely to develop the most aggressive type of brain tumor, glioblastoma, than women, while female patients tend to survive the same disease longer and have cancers that respond better to therapy. Is it genetics? Is it hormones? Or is it something else?

To get to the bottom of the mystery, Swanson teamed up

* Men can get breast cancer as they have a small amount of breast tissue, but it's rare. There are around 2,600 cases every year in the US compared with more than 276,000 female breast cancers.

with pediatric neuroscientist Joshua Rubin and his team at Washington University School of Medicine in St Louis, Missouri. Together, they took an in-depth look at data from thousands of patients, as well as studying brain tumor cells growing in the lab and transplanted into mice. Fascinatingly, they found distinctive patterns of gene activity and response to treatment in male brain tumor cells compared with female ones. This couldn't be explained by the influence of male or female hormones, suggesting that there's a fundamental difference in the genetic programming in cancer cells depending on their sex.

Swanson suspects this goes all the way back through the evolutionary history of the disease. Perhaps, she suggests, cancer cells of different sexes might choose different strategies to survive in the stressful environments induced by treatment. By way of example she points to the different ways in which male and female fetuses respond to food shortages in the uterus. In typical famine situations, such as in sub-Saharan Africa or parts of Europe during World War II, a normal number of female babies are born, but they're unusually small. But for males, the opposite is true: fewer boys are born but they're all a normal size.

This makes sense in terms of evolutionary programming: if only a few strapping males of a species are needed to impregnate a much larger number of smaller females, then it's most efficient to allocate biological resources along these lines when times are tight. In a cancer, the equivalent of a famine would be a stress such as radiotherapy, chemotherapy, or the restricted, messed-up blood flow that's often found in tumors. And therefore, according to Swanson and Rubin's data, the pattern of male cancer cells growing big and strong versus slow-and-steady female cells seems to be recapitulating this evolutionary strategy right down on a cellular level.

The idea that cancer cells might be acting out a deeper evolutionary program depending on the sex of their host is fascinating, if more than a little controversial. Her discovery has big implications for personalized approaches to treating brain tumors—not only should oncologists take the presence of particular driver mutations into account when deciding on a treatment, but they should consider the underlying genetic sex too. It's also intriguing to wonder whether or not the same pattern plays out in other types of tumor, especially given that most cancer drugs are tested on male cancer cell lines and in male animals.

One further question raised by these findings is what happens with brain tumors in transgender or intersex people, particularly those taking hormone therapy. Do their cancer cells adhere to the underlying genetic program encoded in their sex chromosomes, or are there other biological and hormonal factors that come into play? There are very few of these cases around, but Swanson is now doing her best to recruit gender non-conforming and trans patients to see how their brain tumors behave and whether they map back onto the typical pattern for one sex or the other.

As highlighted by the discovery that it may not be possible to separate the behavior of tumor cells from the underlying genetic sex of the person they arose in, it's important to remember that cancer is still part of the body. There's a common misconception that cancers are somehow "other" alien beings that are growing inside us, rather than the product of our own tissues. But they're still cells, however messed up they may be, and they're still going to do the things that cells do. And in the case of brain tumors, this includes wiring themselves directly into their neighboring neurons.

Three intriguing papers published in late 2019 showed that glioma brain cancer cells could form functional electrical connections (synapses) with healthy nerve cells, hijacking normal survival signals to help them to grow and spread. Breast cancer cells that have spread to the brain also appear to wire themselves in (at least in mice). Cancer patients often talk about "chemo brain"—the foggy forgetfulness that's a common side effect of treatment. Maybe there's also such a thing as "onco brain," if these unwanted cancer connections are affecting normal brain functions. That's still very speculative right now, but it's definitely high up on the lengthy list of Really Weird Things I Discovered While Writing This Book. However, the story that follows is probably number one.

## TWO BECOME ONE

As I was busily trawling the literature and interviewing scientists, I'd occasionally hear whispers of something so transgressively bizarre that it made my head spin. Not only could cancer cells evolve any and every imaginable invention in their journey from cellular cheat to invincible tumor—*they were also having sex*. This would be huge, if true. We're used to the idea of cancer cells reproducing asexually by splitting in two, similar to yeast or bacteria. But what if it's possible for them to fuse together, pooling their genetic assets and spawning even more deadly offspring? Just imagine the implications for our current understanding of tumor evolution if cancer cells could pick up and spread resistance mutations between each other, rather than having to go to all the trouble of mutating themselves.

Yet every time I tried to pin down some solid evidence for this rumor I drew a blank. Someone had overheard a guy talking at a conference or in the bar afterwards but couldn't remember the details. I uncovered a handful of relatively obscure reviews in scientific journals suggesting that the unusually large cells with doubled genomes that are occasionally found in tumors could have been formed by two cells fusing together rather than one cell failing to divide properly, as is usually believed. Occasionally, a couple of researchers I interviewed made vague references to strange goings-on in lab-grown cells, while one admitted to trying to force the issue by growing two types of tumor cells together in the same dish and seeing if they would breed.

I'd virtually given up hope of discovering the truth. But then I saw Kenneth Pienta, a smooth-talking urologist from The Johns Hopkins University School of Medicine in Baltimore, Maryland, give a talk at a small meeting in Paris. While searching for an explanation for the rapid emergence of drug resistance in prostate cancer, he noticed some unusually large cancer cells appearing in resistant tumors. Even more strangely, these giant cells had at least double the amount of DNA than might be expected. Normal cells of the body with two sets of chromosomes (twenty-three pairs, one of each from mom and dad) are referred to as diploid, scientifically speaking, while chunky units with multiple genomes are known as polyploid.

To find out more about the origins of these mysterious polyploid giants, Pienta and his colleagues built something they call the Evolution Accelerator. It's a hexagonal microfluidic chip, no bigger than a fingernail, containing a miniature silicon landscape for cancer cells to explore.* Within the chip is a network of tiny interconnected chambers, linked by tunnels that are small enough for diploid cells to navigate but not the larger polyploids. Unlike cells growing in a regular Petri dish, which are all washed with the same levels of nutrients, oxygen, and drugs, the Evolution Accelerator enables the researchers to set up chemical gradients across this tiny world. Pienta and his team set up the chip with low concentrations of a chemotherapy drug (docetaxel) on one side and high levels on the other, threw a bunch of drug-sensitive prostate cancer cells into the arena, and sat back to watch what happened.

Using time-lapse microscopy, the team tracked the cells over several weeks as they meandered around their glass world. Giant polyploid cells quickly began to appear, especially in the areas with the highest levels of docetaxel. The higher the concentration of drug, the larger the number of giants, indicating that they were resistant to the treatment. By

* Imagine the arena in the film *The Hunger Games*, but for cells.

contrast, smaller diploid cells were rapidly killed by the highest concentrations of docetaxel, with any survivors quickly migrating toward less toxic areas.

Taking a closer look, Pienta spotted that there were two ways to make a polyploid. The first was through incomplete cell division, when a cell copies its DNA but doesn't split in two afterwards. This is something to be expected, given that docetaxel works by interfering with the molecular scaffolding that engineers the movements of cell division. But he also spotted a second way to make polyploids: cell fusion. All over the place pairs of diploid cells were joining together to create giant monsters. Then things got really weird. Not only were the cells fusing to make polyploids, but these giant cells were also then birthing new diploid daughters, which were all resistant to the drug.

"They're just popping these cells out," Pienta says, presenting his findings to a stunned audience. "And the more we treat, the more polyploids we get."

Although it sounds outrageous, the discovery of cell fusion inside cancers maybe isn't so weird after all. Fusion happens in other situations in normal human life, such as during the formation of the placenta, when muscle cells fuse together to make long fibers, and in wound healing. The ability to join two cells together is clearly already encoded within our DNA, so it's not so strange that tumor cells can access and activate that pathway.

Researchers have spotted cancer cells completely engulfing whole cells inside tumors (a phenomenon going by the rather charming name of emperipolesis)* and fused cells have previously been found in several types of cancer, turning up in response to chemotherapy, radiotherapy, or changes in the tumor micro-environment. However, they were thought to be a harmless oddity, unlikely to proliferate and probably destined to die.

There's even evidence of cancer cells fusing with healthy cells, which some researchers suspect might be a key trigger for cancer spreading. This idea was first put forward more than 100 years ago by the German pathologist Otto Aichel, who spotted white blood cells attacking cancer cells and wondered if they ever joined forces. Despite intriguing hints from animal studies, it's been hard to pin down whether or not this really happens in humans or if it actually matters. However, a 2018 paper from a team at Oregon Health and Science University in Portland, Oregon presents some compelling evidence for the presence of fusions between tumor and immune cells in pancreatic cancer patients, and that the greater the number of fused cells, the worse the chances of survival.

Now that Pienta knows the giant cells are there and what they're capable of, he's started seeing them everywhere. When he went to look at prostate cancer cells growing in a regular large flask, all bathed in the same liquid environment, he found around 3 percent of the cells were polyploid, with the remainder all being diploid. After adding a hefty dose

* While it's rare under normal conditions, this cellular swallowing also happened way back in the history of life on Earth. One greedy bacterium engulfed another, creating the first complex cell—the precursor of all animals, plants, and fungi alive today.

of docetaxel, the proportion shot up to about 90 percent, then went back down to 3 percent again if the treatment was removed. However, all the diploid cells were now resistant to the drug.

His descriptions of cells fusing and popping out resistant daughters sounds like the elusive phenomenon I've been searching for. At the end of his talk he stops for questions and I cautiously put up my hand.

"What you're describing sounds like sex to me—for varying values of good sex!" I say.

I get a cheap laugh from the audience as Pienta nods in agreement.

"And if that's what's happening then it's . . . problematic?" I continue.

"Yes. And it's scary," he confirms.

He's not joking. The increasing proof that cell fusion happens in cancer has big implications for treatment and the emergence of resistance. Giant cells appear to be rare in primary tumors before treatment has started, but Pienta's results suggest that giving ever higher doses of chemotherapy actually encourages the formation of polyploid drug-resistant cells that appear to act like stem cells, spawning an army of resistant daughters. He also suspects that these large polyploids are the "hardy emigrants" responsible for traveling round the body and establishing secondary cancers.

Pienta shows us a picture of a metastatic prostate cancer that's spread to the lung, made up of a significant proportion of polyploid cells. It also looks like the polyploids can lie dormant until they're reactivated—"They suddenly come out and go BOOM!" as he describes it—potentially explaining why cancers can appear to have been successfully treated and then return with devastating effect at a later date.

As Pienta sees it, these giant polyploids are like the queen bees in a hive—an elite population of cancer cells that are capable of "reproducing" by spawning resistant diploids. They're also the ones that head off in search of a new hive, analogous to metastasis of tumors around the body. Maybe, he suggests, we should be thinking of tumors as super-organisms—groups of individual cells that act together to produce collective behavior. There's no "brain" in a tumor but there might be a "hive mind," emerging as a consequence of the collaboration.

In some ways this is almost as if cheating cancer cells, having rejected their multicellular host and returned to a more unicellular way of life, are starting to team up and reinvent multicellularity again. It's yet another biological invention that has evolved in the past, so it's not entirely unexpected to see it happening within the evolutionary crucible of cancer.

The more scientists investigate the intricate molecular details of cancer, the more weirdness they see. Yet to my mind, none of this should be surprising. It's just evolution, doing its thing. The history of life tells us that evolution can generate incredible diversity. Multicellularity has evolved multiple times. Sex has evolved a few times, too. Species multiply, migrate, adapt, and diversify. Proliferators gonna proliferate. Mutators gonna mutate. Life just keeps shaking it off.

From the smallest bacterium to the biggest blue whale, species have strived, survived, thrived, and died upon the Earth for millions upon millions of years, their genes, cells and bodies shaped by natural selection according to the environment in which they live. As Darwin wrote in his master work, *On the Origin of Species*:

> There is grandeur in this view of life . . . whilst this planet has gone cycling on according to the fixed law of gravity, from so simple a beginning endless forms most beautiful and most wonderful have been, and are being, evolved.

Of course, nobody's arguing that cancer is beautiful and wonderful. It's terrible and ugly and destructive and it steals the people we love. It's an absolute bastard of a disease that indelibly changes everyone who experiences it. But it's also a textbook example of natural selection in action—a dumpster fire of evolution compressed into months or years instead of millennia. This isn't a smart long-term strategy, as cancer often ultimately leads to its own extinction, dying along with the unfortunate body in which it arises.

Given that cancer cells appear to have explored every biological avenue that we know of, the discovery that they can fuse and spawn shouldn't be surprising. On a fundamental biological level, the drive to have one last reproductive roll of the dice when everything's falling apart makes perfect evolutionary sense: if you don't make it, maybe your offspring will. Entire movie plots have been predicated on the idea of seeking solace in the arms of another during times of disaster. Bombarded with toxic drugs, scorched by radiotherapy, and facing an army of immune predators, what should cancer cells do when it seems like their world is ending? Have sex.

Then, according to the movies, they should try to leave their doomed host planet.

So maybe we also shouldn't be surprised that they've occasionally managed to make the ultimate evolutionary leap.

## UNLUCKY DEVILS

Perhaps the most famous four-legged inhabitants of Tasmania, the rugged island off the South Australian coast, are its eponymous devils. These carnivorous mammals are solitary and nocturnal, earning their name owing to their vicious behavior toward their kin, aggressively screeching and biting each other in the face if their paths cross. The black fur, fierce red ears, mean stare, and habit of feeding on dead flesh adds to the satanic allure.* Even their Latin name, *Sarcophilus harrisii* ("Harris' flesh-lover"), has more than a little touch of the darkness about it.

Sadly, the Tasmanian devil's gothic good looks are under threat from Devil Facial Tumor Disease (DFTD)—an aggressive and unpleasant cancer that forms ulcerated tumors around the animals' mouth and jaws, eventually metastasising into their internal organs. Unfortunately for the already endangered devils, the cancer spread rapidly through the population since the first cases were reported in the mid-1990s.

* The German name for the Tasmanian devil is *beutelteufel*, which I originally mistranslated as "the devil's handbag." It actually means "bag devil," the bag referring to the pouch common to all marsupials.

In just a few years it had led to a major collapse in the population, with only a few cancer-free pockets remaining.

When DFTD was first identified, the way it spread through the population made people think it might be owing to a virus, similar to Peyton Rous' chicken sarcomas or Shope's jackalopes (page 84). There were also concerns that the same virus might be jumping into humans, sparked by an unusually high number of blood cancers in rural areas of the island.

The discovery of its true nature fell to Anne-Maree Pearse, a Tasmanian government researcher. Throughout the 1980s, Pearse worked at the Royal Hobart Hospital in Tasmania as a cytogeneticist, specializing in studying the faulty chromosomes in tumors to help with diagnosis and treatment. As well as looking at cancers from human patients, Pearse was also kept busy with a steady stream of samples from devils. By 2004, her interest in the growing scourge of DFTD had led her to become the senior cytogeneticist for the Save the Tasmanian Devil Program in the government's Animal Health Laboratory, with the hope of pinning down the culprit responsible for all these cancers.

Almost immediately she noticed something very odd about the chromosomes in all these tumors: they were all exactly the same. And, even more strangely, they bore no resemblance to the animals from which they'd been taken. This was perplexing in the extreme. Because each cancer arises from cells within that individual's body, every tumor should be a unique genetic event with its own particular chromosomal idiosyncrasies, including cancers caused by infectious viruses. Yet her results looked for all the world like it was the cancer cells themselves that were being transmitted between animals and spreading the disease.

Publishing this curious discovery in a single-page paper in the prestigious journal *Nature* in 2006, Pearse and her departmental colleague Kate Swift put forward the idea that DFTD was an infectious cancer, rather than a virus, that was passing from one devil to another. Further work from a team at the University of Sydney, Australia, conclusively proved Pearse and Swift's hypothesis correct: the facial tumors were caused by an immortal rogue clone of cancer cells that had somehow escaped the bonds of the animal in which it originally arose and become transmissible. Yet nothing was known about exactly where this adventurous cancer had come from and why it had become contagious.

Growing up in Tasmania, geneticist Elizabeth Murchison was used to spotting deceased devils on the side of the road, often knocked over while scavenging on other animal road traffic victims. Now a group leader at the University of Cambridge, UK, she and her team are studying both the origins and the genetics of DFTD, in order to help save the species. One of the first specimens they studied was collected by Murchison herself, who spotted the corpse of an infected devil while driving home from a backpacking holiday and popped it into the back of her car.

In 2010, she published her first major paper about the devil tumors, comparing cancer samples from that first hapless animal with a few others that she'd collected over the years. By comparing the patterns of gene activity between the cancer cells and different body parts of a healthy devil, she realized that the cancer probably started from a Schwann cell. These cells normally act as a kind of electrical insulation tape, wrapping around nerve cells and protecting the electrical signals to and from the brain. Interestingly, they also love to move. Schwann cells migrate rapidly through the body and spread along the long cables of nerves, so they already have a strong tendency to spread. Maybe making the leap between individuals was just the next step in their evolutionary journey.

Murchison and her team also started digging into the DNA of the devil tumors. Many parts of the tumor DNA are highly similar to the genomes of devils alive today, suggesting that the original devil that spawned the cancer lived fairly recently, probably in the late 1980s or early 1990s. Like mammals, female marsupials have two X sex chromosomes, while males have an X and a Y. Although Murchison and her team found no obvious sex chromosomes, they found what looked like the remnants of two X chromosomes embedded elsewhere in the tumor genome and no trace of a Y, suggesting that the first founder devil was probably female. Although she died before the disease started to become obvious across the island, her cancer cells live on, continually evolving and changing as they spread through the rapidly shrinking population.

It was odd enough to have found one contagious cancer that had evolved so recently with such devastating effects. So it was doubly strange when Murchison and her team found a second. While analyzing tumor samples from five devils found in Southern Tasmania, the researchers were stunned to see that the chromosomes in the cancer cells looked completely different from the original DFTD cells yet were all the same as each other. The big giveaway was the presence of a Y chromosome, proving that the founder of this second tumor must have been male. On the surface, however, it's impossible to tell the difference by looking at infected animals or tumors. To misquote the Irish poet Oscar Wilde, to have one transmissible cancer in your species may be regarded as a misfortune; to have two looks like carelessness. So what's going on?

"Two things need to happen for a cancer to become transmissible," Elizabeth explains to me, as we sit in her office decorated with several stuffed toy devils. "First, it has to find a means of escape from one host to another and secondly, it has to acquire adaptations to evade the immune system that would see it as a foreign graft. Either of these things is unlikely to happen and it's even less likely that they would happen together."

Looking at how these normally solitary animals act toward their neighbors, it becomes easy to see how DFTD manages to check the first box. Although they're quite docile around humans, devils don't play nicely with each other. As they fight and bite, clumps of cancer cells are torn off the jaws of an infected animal and lodge in the fresh wounds that it has inflicted on its opponent. Without such an easy route of transmission, it's unlikely that DFTD would have gained such a lethal foothold in the population. So that leaves us with the second challenge: Why doesn't the devil's immune system recognize and reject this invader?

Mammals and marsupials have evolved highly complex

immune systems that continually seek out and destroy anything that looks like it doesn't belong in there, including cells from a stranger, whether of the same or another species. The task of spotting the difference between "this is me" and "this is not me" is made easier by genes known as the major histocompatibility complex (MHC). These are the most diverse parts of the genome, encoding molecules that stick out of the surface of cells like waving flags. If these flags look strange or foreign, then the immune system springs into action and destroys the invader.*

The protective abilities of the MHC system explain why it's so important to match transplant donors with recipients carefully. Even with the best matches, people getting organ transplants have to take immune-suppressing drugs to prevent rejection. Curiously, Tasmanian devils will normally reject tissue grafts from other devils, so they have at least some level of immune surveillance working. It turns out that cells from the first DFTD have completely lost all of their MHC genes, so they can jump into any devil. The second, more recent, cancer still has its MHC genes, but because the Tasmanian devil population is so small and inbred, a number of the animals on the island share them too.

This lack of diversity means that the cancer cells can move between a limited range of genetically similar animals without alerting the immune system. It also looks like this second tumor is on the way to losing its MHC genes altogether,

* There's a lot more about this fascinating area of science in Daniel Davis' book *The Compatibility Gene*.

providing an important clue as to how these cancers are likely to evolve: losing MHC isn't essential for a tumor to become transmissible, but it does increase the capacity for cells to move into a larger population of hosts.

A couple of years ago it looked like DFTD was going to be the end of the Tasmanian devil in the wild. The cancers don't respond to chemotherapy, even if infected animals can be captured and treated in time, and some populations had plunged by 90 percent. Save for a few quarantined "insurance" populations, the rapid spread of the disease and the discovery of a second strain seemed to be spelling disaster for these iconic marsupials. But while we can blame evolution for the adaptations that enabled the facial tumors to make the leap from one individual to another, it may also be helping the devils to fight back.

Wildlife ecologist Rodrigo Hamede and his colleagues at the University of Tasmania have been keeping a close eye on the dwindling devil population, aided by locals who can report devil sightings through a smartphone app. It looks like some of the devils are evolving immunity to DFTD, resisting infection with the tumor cells. They've also found more than twenty cases where infected devils have managed to heal themselves, with gory tumors showing complete regression without any human intervention. It's too early to say whether the devils have managed to save themselves—and a new system of roadside fences and warning signals is helping to reduce the toll of road deaths—but their future looks less imperiled than it did a few years ago.

It's relatively simple to understand the emergence of two

strains of transmissible devil tumors by invoking the animals' unique circumstances: an easy route of contagion coupled with a small population with low genetic diversity. But it's harder to explain how a cancer affecting a much more widespread and genetically diverse species has also made the leap to independence.

## A DOG'S LIFE

Dog sex is not a romantic business. Once the male has ejaculated, the end of his penis swells up and sticks inside the female reproductive tract, keeping the two participants locked together at the point where they'd probably much rather be getting on with their day. Trying to get out of this "copulatory tie" before the swelling has gone down can be traumatic, inflicting damage on the relevant bits of either animal. As with the devils, this behavior has inadvertently led to the emergence of another contagious cancer.

In 1876, a Russian vet named Mistislav Nowinsky noticed an unpleasant-looking cancer affecting the genitals of dogs, which seemed to be passed on by mating. To prove his theory, he rubbed bits of tumor from one infected dog into small cuts that he'd made on the genitals of another, which then went on to get the disease. At the time, this sparked a lively debate among scientists, who were still arguing heatedly about the causes of cancer. The idea that cancer itself might be contagious was intriguing, as well as providing justification for stigmatizing and isolating patients.

Quarter of a century after Nowinsky's experiments,

German vet-turned-doctor Anton Sticker set to work in his lab in Frankfurt to investigate this unusual canine cancer, as well as the potential for other types of cancer (including human tumors) to be transplanted from one individual to another. He confirmed Nowinsky's observations that the tumors could be passed from dog to dog, and the disease even bears his name—Sticker's sarcoma—throughout much of the scientific literature. Others took up the transmissible tumor baton with Alfred Karlson and Frank Mann, two scientists at the University of Minnesota in Rochester, demonstrating an impressive commitment to the cause. Between them, they transplanted the same cancer into forty generations of dogs, finally publishing their findings in the early 1950s.

Regardless, cancer researchers throughout the early twentieth century settled on the idea that the disease was either the result of alterations in chromosomes or the work of infectious viruses. Sticker's contagious sarcoma, now known as CTVT (canine transmissible venereal tumor) was viewed as little more than a scientific curiosity, despite affecting countless dogs in countries all over the world. The idea that tumors could be spread by cells passing from one animal to another seemed bizarre and unbelievable. Many thought that there must be a virus at work, even though attempts to induce cancers with highly filtered cell-free extracts from tumors—exactly the same approach that Rous and Shope had used to find their eponymous viruses (page 84)—had drawn a blank.

This conundrum intrigued Robin Weiss, a virologist at UCL, who felt sure that CTVT must be caused by an elusive virus that nobody had yet managed to find. Determined to be the one to track it down, he set to work analyzing DNA from tumor samples retrieved from sixteen dogs in Italy, India, and Kenya in search of leads. Rather than finding hidden traces of an unknown virus, he discovered the same thing that had baffled Anne-Maree Pearse as she studied her devil cancers Down Under. The genomes of the tumors were all pretty much the same, yet completely different from the animals in which they'd been growing. Forty more near-identical tumors gathered from five continents proved the uncomfortable truth: they were dealing with contagious cells, not an infectious virus.

Weiss's paper was published in 2006, just six months after Pearse's publication about the devil disease. Like devil facial tumors, transmitted through the wounds caused by face-biting, CTVT has spread through the population by exploiting the genital damage inflicted during dog mating. But unlike DFTD, which seems to have sprung up in the past few decades, these dog tumors are much older. By comparing DNA from the tumor with various breeds around the world, the UCL team concluded that the original founder was likely to have been an ancient Asian dog breed from China or Siberia, or possibly a wolf.

CTVT is the longest-lived cancer that we currently know of, picking up a staggering 19 million mutations along the way, and it continues to evolve and adapt in different parts of the world. Like the original devil cancer, the dog venereal tumor cells have lost their MHC "compatibility" genes, explaining why they have no trouble jumping from host to host.

In the same way that Charles Swanton and his team can reassemble the "family tree" of a lung cancer evolving and spreading inside a patient (page 166), Elizabeth Murchison has been able to retrace the route of CTVT as it spread around the world. The latest analysis suggests that the disease first arose in Central Asia somewhere between 4,000 and 8,500 years ago, staying in that area for several thousand years. It started spreading from the first century onward, hitching a ride to the Americas during the sixteenth century as adventuring sailors took their mutts to sea and making a return journey a century later. Murchison has also been able to create a genetic "photofit" for the founder dog in which the tumor first grew, suggesting that it was probably a medium- or large-sized animal similar to today's Alaskan Malamutes, with a black or dirty sand-colored coat, pricked-up ears, and a pointy nose. Unfortunately, she wasn't able to discern if the original animal was male or female, so we have no way of knowing whether it was a Very Good Boy or a Very Good Girl.

Since emerging in this first furry host, CTVT has spread almost everywhere that dogs can be found, with one notable exception. In the summer of 2018, Murchison persuaded Alex Cagan, a young post-doctoral researcher at the Sanger Institute, to take a trip to Pripyat in north-eastern Ukraine—the site of the former Chernobyl nuclear power plant. In April 1986, the plant's Number 4 reactor exploded, showering the area with radioactive fallout. Fleeing local families had to leave everything behind in the rush to evacuate, including many beloved pets. Thirty years later, and the descendants of these abandoned dogs are happily roaming around the place and doing what feral dogs do best: feeding, fighting, and . . . mating.

Given that every other population of dogs on the planet is affected by CTVT, it was a reasonable assumption that these animals would have it, too. Murchison and Cagan were curious to know whether or not the high levels of radiation around Chernobyl had left their mutational mark in the DNA of the cancer cells, which would have been propagated as the disease continued to spread from dog to dog. Teaming up with the Clean Futures Fund, who were running a canine health and neutering program around Chernobyl, Cagan headed into the exclusion zone around the ruined reactor in search of genital cancers (and, judging by his photos on the Sanger Institute's blog, plenty of cute snoots to boop). Yet he returned to Cambridge empty-handed. After inspecting 200 dogs over two weeks, he couldn't find a single transmissible tumor, even though the disease is present in animals living just 90 miles away in the Ukrainian capital, Kiev.

Nobody knows why the pups of Pripyat are free of CTVT. It may just be down to chance; if none of the dogs in the original population of pets had the disease and no strays brought it in from outside, then it's never going to turn up in such an isolated community. Maybe their immune systems are unusually strong, enabling them to overcome the cancer. Or—and this is a really wild idea—the radioactivity in the area has acted as a kind of accidental radiotherapy, effectively treating all the animals in the vicinity and wiping out the disease.

CTVT is highly sensitive to radiotherapy and

chemotherapies that damage DNA, so it's certainly a possibility. But because researchers working on the site over the past thirty years have been rightfully more interested in ensuring human health and radiation security than dog genitals, we may never know.

## FROM CLAMS TO CANNIBAL HAMSTERS

Until 2015, it was thought that CTVT and two strains of DFTD were an outlying trio of intriguing anomalies in the annals of cancer research. Then came the clams.

Since the 1970s, marine biologists have been concerned about a strange disease sweeping through colonies of soft-shell clams along the north-eastern seaboard of the United States. Similar to leukemia in humans, it causes rampant proliferation of haemocytes—the clam equivalent of red blood cells—which clog up its body and eventually lead to death. Up to 90 percent of a colony can die from the disease, which isn't only a biological disaster but an economic one too, as the clams are an important part of the seafood industry.

News of this clam calamity reached Michael Metzger, a young researcher at Columbia University in New York. Like Weiss' quest for the virus behind CTVT, Metzger suspected that a virus might be responsible for the disease and set about trying to find it. By analyzing DNA from infected animals, he found the cancer cells did indeed contain a virus-like piece of DNA called Steamer, which embeds itself in random locations within the genome as part of its life cycle.

Weirdly, the site where Steamer had chosen to muscle into the clam genome was exactly the same in every single cancer sample he looked at, even if they came from clams in completely different locations. It was too much of a coincidence to expect that a mobile element like Steamer would randomly hop into exactly the same spot in clams. After further genetic analysis he came to the weird but inevitable conclusion that this must be yet another contagious cancer, transmitted through leukemia cells pumped into the surrounding seawater by infected individuals.

The North American shellfish beds aren't the only places affected by these kinds of cancers. Metzger wanted to know whether similar destructive diseases in other locations were also caused by the same clam cancer cells drifting around in the Atlantic. Amazingly, he discovered four more completely different transmissible leukemias: one in Canadian mussels, along with two separate cockle cancers and one in golden carpet shell clams off the Spanish coast. Even more weirdly, this last example appears to have originally come from a completely different species, the pullet shell clam. Curiously, pullet shells don't show any signs of being susceptible to the golden carpet shell disease and must have somehow developed resistance along the way. More recently, Metzger and his team have discovered two more types of mussel that are infected with the same transmissible cancer cells, which first arose in a third species. One of these mussels lives in the seas surrounding South America while the other dwells in European waters, suggesting that the cancer cells have somehow made it all the way across the Atlantic in search of new hosts.

Along with the discovery of the second devil tumor, Metzger's startling results took the number of known transmissible cancers that have arisen spontaneously in the wild from two to nearly ten within a few short years, and I wouldn't be at all surprised if that number goes up in the future. However, sifting through the scientific literature reveals several more disturbing examples of cancers that have crossed the barrier from one individual to another, even if they haven't become contagious on a widespread level.

Most of these examples come from pregnancy, as the intertwined blood supplies in the placenta provide a handy conduit for wandering cells. Around twenty-six examples of cancer passing from mother to child have been reported in more than 150 years, mostly melanoma or blood cancers. Given that there are more than 100 million babies born every year, and around 500,000 or so mothers might be expected to have cancer (whether diagnosed or not), the chances of this happening are extremely low.

Cancer cells can also pass between identical twins in the uterus. The first case of identical twins suffering from childhood leukemia at the same time was recorded in Germany in 1882, and more than seventy examples have turned up since then. Careful genetic analysis has shown that these cancers must be the result of a clone of rogue cells arising in one twin, which then spreads to its sibling through the knotted blood vessels of their shared placenta. There are also rare cancers known as choriocarcinomas that start in the placenta—a tissue that starts out as part of the early embryo—and can spread into the mother's body.

Then there are artificial routes of transmission. In March 2018, doctors in the Netherlands published an extraordinary report about four people who had all developed cancer after receiving organ transplants from the same donor—a fifty-three-year-old woman who had died from a brain hemorrhage. At the time of her death, there was no sign that anything was wrong and certainly no obvious tumors. Yet the three recipients of her lungs, liver, and left kidney all died from metastatic breast cancer within seven years of their transplants. Alarmingly, all three cancers appeared to be identical.

The fourth, a young man who got the woman's right kidney, also developed cancer, which was only successfully treated by removing the donated organ and taking him off the immunosuppressive drugs that are essential to prevent the recipient's body from rejecting the transplant. In this case, it was the right thing to do: his immune system sprang into action and wiped out the tumor cells. As of April 2017, he was apparently completely cured and on the waiting list for a second, luckier organ.*

It's also possible to catch a transmissible cancer by accident. There's the example of a surgeon who nicked his hand while removing a malignant abdominal tumor from a young man. Five months later, a tumor the size of a golf ball had sprung up on his palm exactly where he'd been cut, with

* It's worth remembering that the chances of "catching" cancer from a transplant are incredibly rare—less than one in 2,000—and certainly much smaller than the risk of dying while waiting for a desperately needed transplant. Please join your country's organ donor register.

tests proving that it was the same cancer that he'd previously removed from the belly of his patient. Then there's the unlucky laboratory worker who was carrying out experiments involving injecting human bowel cancer cells into mice and accidentally poked the needle into her left hand. Within a fortnight there was a small lump, which turned out to be made of the very same tumor cells she'd been working with. Although it was successfully removed and she suffered no ill effects afterwards, it's a salutary lesson for anyone who, like me, has been both clumsy and cavalier in the lab.

And then there are the deliberate ones. In a story straight from the Annals of Absolutely Terrible Research That Should Never Have Been Allowed, New York-based oncologist Chester Southam spent much of the 1950s and 1960s injecting people with cancer cells without their proper consent. Some of the recipients were cancer patients who had come to see him in the hope of a cure. Others were elderly people with dementia who'd been admitted to Brooklyn's Jewish Chronic Disease Hospital in New York, while the rest were otherwise perfectly healthy convicts at the Ohio State Penitentiary, many of whom were black.

Not only did Southam pick on people who weren't able to consent properly to his experiments, whether through desperation, neurodegeneration, or incarceration, he didn't even tell the recipients what they were getting. Instead, he preferred to refer to "human cells grown in the lab" so as not to freak people out about the fact they were about to be injected with cancer.

Working together with his colleague at New York's Sloan

Kettering Institute, virologist Alice Moore, Southam showed that healthy people's immune systems would quickly reject transplanted cancer cells in a matter of weeks, without exception. But patients with advanced cancer took much longer to respond and in some cases the injected cells grew steadily into new tumors over a period of months. Despite being told that they weren't in any danger, two of the patients died unexpectedly and four had to have these new tumors surgically removed. Occasionally the cancer came back and in one case the disease started to spread through the body.

Southam's approach was deeply unethical and sparked an outcry among his fellow doctors. It remains a very dark spot in Sloan Kettering's story and the wider history of cancer research. Yet there was method in his seeming madness. As an immunologist, he was interested in the potential for using "foreign" cancer cells to awake a patient's immune system to attack their disease. He wasn't the only person following this approach, as revealed by a particularly sad story published in 1964 by researchers from Northwestern University in Illinois. In 1958, a fifty-year-old woman had an operation to remove a melanoma skin cancer that had appeared on her back. By 1961 it was back with a vengeance and she was given chemotherapy along with a blood transfusion from a patient who'd successfully been treated for melanoma a few years previously.

Realizing the gravity of the situation, the woman's eighty-year-old mother agreed to be injected with her daughter's cancer cells in the hope of generating antibodies against the tumors that were now ravaging her child's body. The mother was in good health on August 15, 1961 when doctors transplanted a half-centimeter chunk of her daughter's melanoma into her abdominal muscle. Tragically, there was no hope of this last-ditch attempt yielding a cure, as the daughter suddenly died the very next day from a perforated bowel.

Around three weeks after the transplant, her mother began to complain of an uncomfortable "pulling sensation" in her tummy. It was clear that the melanoma had already begun to grow, so it was swiftly removed, along with significant chunks of her muscle and skin. Despite this drastic intervention, the woman quickly developed metastatic melanoma and died less than fifteen months after the transplant, her body riddled with the same tumors that had previously consumed her daughter.

All these stories add up to a picture of cancer transmission between humans being possible, if incredibly rare. But there's one more tale that highlights just how weird things can get. Early in 2013, a 41-year-old man walked into a clinic in Medellín, a large city high up in the Colombian mountains. Diagnosed with HIV seven years earlier, he was in a bad way. He'd been skipping his treatment, was losing weight and coughing persistently, and was consumed by fever and exhaustion.

The most obvious diagnosis was worms, based on the presence of parasitic tapeworm eggs in his stools. But he also had strange nodules in his lungs, liver, lymph nodes, and adrenal glands. A course of worming tablets didn't do much to help and the lumps continued to grow. When he came back a few months later the doctors took a closer look at these unusual growths. Although they looked like tumors—packed with proliferating cells, plumbed with blood vessels, and invading into neighboring tissue—there was something very odd about the cells themselves. They were much smaller than typical human cancer cells, yet they didn't look like normal tapeworm cells or any other kind of parasite.

After sending samples up to the US Centers for Disease Control and Prevention (CDC) in Atlanta, Georgia, the horrifying truth emerged: the tumors were made of tapeworm cancer cells. The disease had presumably originally developed in one of the parasites infecting the man's gut and invaded the rest of this body, and his HIV-damaged immune system was powerless to stop it.

The CDC team reported their grisly findings back to the man's doctors in Colombia, but they were too late. He was already very sick owing to the combined complications of advanced HIV and the tapeworm tumors, and he died three days later. As far as we know, this is the only example of a cancer managing to jump the species barrier between parasite and human. But given the relatively high prevalence of both tapeworms and HIV in many parts of the world, combined with relatively poor cancer diagnostics and data collection, we have no idea whether or not this was a one-off.

Rather than being a biological curiosity, transmissible tumors are starting to look like they might be A Thing, after all. The big question is how rare are they? The unifying factor that pops out of all these stories of contagious cancers is that failure of the immune system seems to be critical for allowing transmissible cancers to gain a foothold in the body.

The unlucky transplant recipients in the Netherlands were all taking immunosuppressive drugs, while the immune system of the man with the tapeworm tumor had been trashed by HIV. The dog and devil cancers have all found ways of evading immune detection by manipulating the MHC system. Although Chester Southam's experiments were ethically terrible, they proved that a healthy, functioning immune system is usually capable of fighting off invading cancer cells (although the stories of the clumsy surgeon and lab technician would suggest that this isn't always the case).

In fact, there's even an idea that the whole system of immune recognition and rejection evolved in the first place to protect animals against transmissible cancers. And, more controversially, some researchers have even suggested that sex might even have evolved partly in order to suppress transmissible tumors. The random shuffling and dealing of genetic variations involved in making eggs and sperm makes it likely that your cells are different from the other individuals around you, even if you're closely related, making it less likely that a cancer might transfer from one to another.

Given that all the examples of cancer cells transmitted between humans so far have been under unusual circumstances, I'm curious whether or not a true contagious cancer could ever turn up in our own species. Given that the most obvious route for transmission would probably be through sex, many of us would hopefully be canny enough to spot something amiss "down there"; although the persistence of unpleasant sexually transmitted infections suggests maybe not. Direct contact doesn't have to be the only route for living cells, though. Mosquitoes are responsible for transmitting malaria parasites responsible for more than a million deaths every year. So could they also pass on cancer cells?

Horrifyingly, the answer turns out to be yes. Back in the 1960s, an unusual contagious cancer arose in a laboratory colony of hamsters. The origins of the disease were unknown, but it quickly became obvious that it was being transmitted by cannibalism. In order to stop the spread, the animals were kept separated with wire mesh barriers in their cages. Yet the tumors continued to appear, growing from cancer cells spread by coughs and sneezes. And under carefully controlled experimental conditions, researchers even managed to show that mosquitoes could transmit the tumors. It's an artificial lab system and a highly contagious disease, but the tale of the cannibal hamster cancer tells us that insect-borne spread is at least possible.

Transmissible tumors are undeniably rare, but they demonstrate the evolutionary capacity of cancer cells to exploit new environments and evade the protective powers of the immune system. By reshuffling the genetic deck to generate novel possibilities—even to the point of transcending the life of the organism in which they first arose—cancer is a powerful, deadly example of evolution in action, which is exactly why it's so hard to treat successfully.

# 9

# THE DRUGS DON'T WORK

Just before Christmas 2015, British IT consultant Crispian Jago was admitted to hospital for surgery to remove a massive kidney tumor that had already begun to spread into his liver. The operation appeared to be a success, but by the following summer it was back. This time, the outlook was bleak. It's always difficult to make exact predictions about this kind of thing, but he was told that he probably had about eighteen months to live. Remarkably, he's still alive when I catch up with him nearly four years after his initial diagnosis. But rather than owing this extra life to modern genomic sequencing and molecularly targeted therapies, he's never had his tumor DNA analyzed. Instead, he's being kept alive by the informed guesswork of his oncologist at University Hospital Southampton, Matthew Wheater.

First, Crispian tried a drug called Votrient (pazopanib). This initially seemed to work, shrinking his tumors by around 10 percent after three months. But his luck eventually ran out and the cancer evolved resistance. By the summer of 2017, tumors were continuing to grow and spread throughout his body, and his options were running out.

He was given a new immunotherapy drug called Opdivo (nivolumab)—a treatment that works for some but not all of the patients who take it. He wasn't to be one of the lucky ones and just two months into the treatment it was clear it wasn't helping at all. But, as it turned out, although Votrient had only kept his cancer under control for a year, it was long enough to get him to the next new thing.

After ditching Opdivo, Dr Wheater switched Crispian over to Cometriq (cabozantinib)—a new drug that had only just been approved on the NHS. Within a week of starting treatment he was feeling better and within a few months 95 percent of the cancer had gone. To the amazement of his doctors, the drug seems to be holding the cancer at bay, although everything we know about tumor evolution tells us resistance will inevitably emerge at some point. Even so, Crispian remains optimistic that something will turn up.

"I suppose I'm right at the cutting edge," he shrugs. "I've just got to keep going long enough for the next drug to come in."

I don't think there's a standard way that someone with terminal cancer should look or behave or be, but he seems so well and full of life that it's hard to imagine that his body is gradually being taken over by selfish cells. The main change I've noticed has been seeing his hair and beard change from dark brown to shocking white within a year. He's still rocking the three-piece tweed suits, even in the height of summer, and is busy buying every Pink Floyd album he can get his hands on. If I can approach adversity in my life with even a fraction of the positivity, stoicism, and good humor that he's managed over the past few years then I'll be happy indeed.

"At the time they told me it had spread and was inoperable my daughter, Indie, was in her first year at university, so she was two years away from graduation," he says. "They'd told me I had eight months, so I thought *Well, I'm probably not going to make it*, but it was close enough to have as a target."

Not only did he get to see Indie graduate with first-class honors, but he's also feeling confident that he might stick around to see his younger son, Peter, receive his degree in 2020. Given that he's just seen in the first year of the twenties, I'd say the odds are pretty good.

As well as trying to stay alive for his children's graduation ceremonies, Crispian had another target to aim for along the way: outlasting his aged Labrador, Wilbert. The dog's luck sadly ran out in October 2018 and he's now buried in the garden of the charming country cottage where Crispian lives with his wife, Tori. At the time of writing, he's feeling well and enjoying spending time playing with Wilbert's replacement: a black Lab puppy named Stanley.

But there's also less good news. A scan in the summer of 2019 revealed a tumor growing in the front of his brain. Although his treatment is keeping the cancer in check from the neck down, it can't pass through the barrier between the brain and the bloodstream, unlike sneaky cancer cells. It's clearly a massive bummer, but he's more annoyed by having to surrender the keys to his beloved Porsche than about having radiotherapy to control this unwelcome new addition.

## BEYOND WHACK-A-MOLE

By the time it becomes obvious enough to be diagnosed with current techniques, a typical tumor will contain anywhere between a billion and a trillion cells, each of which probably has tens of thousands of genetic mutations and alterations. It will already be a complex ecosystem packed with different species of cells living and dying in diverse microhabitats—some accustomed to the suffocating toxic swamp, others preferring more pleasant conditions.

Every individual person's disease is a special snowflake, unique to their original genetic makeup and the evolutionary processes that have forged it along the way. And once it has started to spread beyond a certain point, resistance and relapse are virtually inevitable. There will be some cells in there, somewhere, that can withstand everything in our medical arsenal. And, ironically, the more targeted and specific a drug is, the easier it is for a cancer to evolve its way around it.

Crispian's surprising survival is a testament to a century or more of dedicated research. It's the kind of success story that warms the cockles of charity fundraisers' hearts and swells the coffers of pharmaceutical companies. But it's also a classic example of the game of biological whack-a-mole that modern oncology has now become: try a treatment, wait for it to stop working, try another one. Rinse and repeat until you're out of options. That may come sooner or later, depending on the type of tumor and the treatments available.

Cancer treatment is increasingly being driven by the concept of precision oncology. Originally, this used to mean using drugs that target specific faulty molecules in cancer cells—like Glivec for leukemia driven by the Philadelphia chromosome or Herceptin (trastuzumab), which targets breast cancer cells carrying extra copies of a cancer driver called HER2—along with a diagnostic test to determine which patients might be suitable.

The definition is sometimes expanded to include any targeted therapy designed to block specific signals within cancer cells, which would include Crispian's treatments, pazopanib and cabozantinib—both kinase inhibitors that block several different proliferative signals in cancer cells. These are sometimes referred to as "smart" drugs, as opposed to "stupid" conventional chemotherapy.

Oncologists are getting used to the idea of selecting a treatment based on the presence of a particular faulty gene or molecule—an "actionable mutation"—regardless of where in the body it turns up. It's less important whether a tumor is in the bladder, bowel, or breast: what really matters is whether or not the cells have a mutation that can be targeted with a drug. Until recently, the high costs of genetic testing meant that this approach could only be applied to limited panels of the usual suspects. As DNA sequencing technology has become ever quicker and cheaper, the ability to sequence the whole genome of a tumor in search of actionable mutations will start to become mainstream.

The concept of picking a magic bullet based on targeting

the specific driver mutations in an individual patient's tumor is rapidly moving from fantasy to reality. It feels fantastically futuristic and it fits with our modern sensibilities to give each person a personalized therapy that's precisely chosen for their specific tumor, rather than a one-size-fits-all standardized protocol. This paradigm of precision oncology has become an article of faith in the world of cancer research—incontrovertible proof that three decades of cataloging cancer genes and developing clever (and very expensive) drugs to target them have been worth the effort.

There's plenty of excitement about the potential for this approach to transform survival for people with advanced metastatic cancer, but the reality arguably doesn't live up to expectations. Right now, the truth is that most people's cancers aren't tested or don't have a suitable genetic alteration for one of these magic bullets.

Oncologist Vinay Prasad and his colleagues at Oregon Health and Science University in Portland, Oregon took a look at around thirty targeted drugs that had been approved by the US Food and Drug Administration (FDA) for use in combination with genetic testing of a patient's tumor since 2006. Out of the half a million or more people diagnosed with metastatic cancer in the USA twelve years ago, around 5 percent (one in twenty) would have been eligible for one of these treatments. By 2018, this number had increased to just over 8 percent. Bear in mind, that's just patients who would have been eligible had their cancer been tested. There are plenty of patients who never get their tumors analyzed at a genetic level, for reasons of cost and practicality. And there's no guarantee that a national health service or medical insurance would be prepared to pay for these often eye-wateringly expensive therapies, even if a genetic test did suggest they might be effective.

Even worse, there's little to show in the way of benefits: Prasad estimates that just over half of the already small fraction of patients who'd have been suitable for targeted therapy selected by genetic testing would have seen any benefit, with an average response lasting just under two and a half years. Overall, the number of people eligible for gene-targeted therapies is edging up by around half a percentage point every year. While it's better than nothing, it's certainly not the Quantum Leap in cancer therapy that you might believe from the media.

Headlines talk about Holy Grails, game changers, miracles, and marvels. We're told that these new therapies are transformative home runs, finally bringing the cures that we've sought for so long. Without wishing to sound too pessimistic—or to dismiss the progress that has been made in prolonging survival in recent years—the true picture isn't quite so rosy.

In another study, Prasad looked at news stories that used these kinds of superlatives to describe new cancer drugs and found that half of the therapies hadn't been approved by the FDA, while one in seven had only been tested in laboratory studies and hadn't ever been near a human patient. This hyperbole is mostly the work of overexcited journalists, but doctors, industry experts, patients, and politicians are also guilty of spinning overinflated stories of success.

The increasingly mainstream use of genetic testing to

determine treatment presents its own challenges. In 2017, a team at the University of Washington in Seattle posted tumor samples from nine cancer patients to two different companies offering the latest DNA sequencing technology for detecting mutations. The results should be concerning to anyone who thinks that this cutting-edge precision oncology is ready for prime time.

One patient had no detectable genetic alterations, according to both providers. For the remaining eight, only a fifth of the mutations detected in their samples were the same from both platforms. And when the companies made recommendations for targeted drugs based on these results, five patients were recommended completely different sets of therapies. Discounting any technical errors or discrepancies between the two tests, this result still shouldn't be surprising. We already know that a typical tumor is a patchwork of genetically distinct clones, and that the mutations that are found within it will depend on which bit has been sliced off and sent for testing.

There are other problems for this paradigm of precision medicine. For example, finding an actionable mutation in a tumor that can be targeted with a drug doesn't necessarily mean that it will work. Researchers are now discovering that drugs designed to target the faulty genes that repeatedly turn up in many different cancers are effective in some tumor types but useless in others, even though they have the "right" mutations.

As an example, Zelboraf (vemurafenib)—the poster child for targeted therapy—is designed to hit an overactive proliferation signal caused by a particular mutation in the *BRAF* gene. It helps to extend the lives of people with malignant melanoma whose tumors contain the faulty gene, but it does nothing for bowel cancer patients with exactly the same mutation. The tumor cells quickly "rewire" their internal pathways, activating an alternative signal that keeps them proliferating as fast as before.

There's a common idea that modern "smart drugs" are better than "harsh chemo," because they supposedly have fewer side effects. This is based on conventional cycles of intravenous chemotherapy with a few weeks off between treatments, where someone might experience grim side effects for a few days and then get a break. Surely, you might think, these clever new therapies should lead to a better quality of life for patients. But that's not necessarily the case.

A study looking at the results of thirty-eight clinical trials of new drugs involving nearly 14,000 people with twelve different tumor types failed to find any significant correlation between the length of survival and something known as "health-related quality of life." This is a measure of physical, emotional, and social wellbeing, as well as the impact on work or other tasks. Overall, the new drugs staved off the time taken for the cancer to return by an extra 1.9 months on average, compared with patients receiving a control treatment.

The side effects that Crispian experienced with his first course of pazopanib were bad enough to land him in hospital. That's unusual, but it's not unheard of. Furthermore, many targeted therapies are now designed to be taken as daily pills, with daily side effects that can quickly become debilitating. As an example, what doctors would describe as "grade 3/4 diarrhea" is classed as having seven stools per day over what's normal, yet it's described in clinical trials as "tolerable." Planning every day around toilet stops may be a reasonable price to pay for a drug that's going to save your life, but many of these therapies provide modest survival benefits at best.

It's this last issue that nobody really wants to talk about, but it's the biggest problem of all. The headlines may be impressive, but the exciting new drugs that we read about in the media aren't cures. They're not even close. While we've got pretty good at treating early stage cancers, particularly in wealthier countries, survival from late-stage metastatic disease is still usually measured in months or single-digit years (although there will always be exceptions).

In 2014, Professor Tito Fojo from the US National Cancer Institute looked at more than seventy new cancer drugs that had come onto the market between 2002 and 2004, all of which cost many thousands of dollars per year. For all their fanfare, each of these fancy new pills increased average survival by just two months. A subsequent report from another team pushed that up to just under three and a half months, although that may be an overestimate partly as a result of focusing on results from short-term clinical trials rather than those with longer follow-up periods.

There are occasional miracles—the young mom or much-loved granddad who's told they have "just months to live" and goes on to "defy their doctors" thanks to a new "wonder drug." Or people like Crispian, still flipping his prog rock records over to the B-side and flipping the bird to cancer. Nobody is truly "average"—and extra time really does matter—but the reality is that most of these drugs extend survival by a matter of months compared with conventional treatment. Furthermore, they are now some of the priciest substances on Earth. A typical targeted therapy costs six times more than plutonium, weight for weight, while the latest CAR-T immunotherapy clocks in at over a billion dollars per gram. Yet, there's very little correlation between the cost of a drug and its effectiveness or the amount of extra life it brings.

So we have to ask why so many of these new therapies are approved at a cost of tens or hundreds of thousands of dollars per year, unlocking millions in profits for the companies that make them, when they bring such limited returns?

The answer to this question partly lies in the way that clinical trials that are carried out in order to gather the data used by regulatory agencies in deciding whether or not to approve a new treatment. Many trials use something known as "progression-free survival" as a measure of how well a drug works—how long it takes before tumors have grown back bigger than they were before the start of the treatment. Far too few focus on overall survival: Does this expensive and potentially unpleasant drug genuinely prolong life? While a new therapy might appear to make an impressive difference in terms of progression-free survival by holding the disease in check for longer than an older treatment, it might not actually lead to an increase in overall survival if there's a short, sharp relapse at the end.

Another sleight of hand is to use what's known as a "surrogate endpoint," such as the level of a particular molecule in the blood that tracks in step with tumor size. While these markers can be useful for revealing whether or not a patient is heading in the right direction, they still don't provide a hard answer to the question that all patients really want to ask: Will this drug give me more time?

We also might expect that new drugs should be tested against the very best options currently available. But that's not necessarily the case. There are examples of therapies being tested against treatments that are no longer considered to be the best standard of care. Some trials compare survival for people taking a new drug today with historical survival times that may no longer be accurate. Many fast-track drug approvals are often made on the basis of initial progression-free survival or surrogate endpoint data against these "straw man controls," in the hope that companies might follow up with longer-term overall survival figures at a later date. I leave you to guess how often that happens.

What's more, the patients who take part in trials tend to be on the younger side, relatively fit, and without major health problems (other than cancer, obviously). They're usually highly motivated to take part, monitored on a regular basis, and are more likely to adhere to the treatment. The idealistic setting of a clinical trial is no match for reality: most people with cancer are often older and sicker than typical trial populations. There are all sorts of additional health issues that might limit the options available for treatment or the doses that can be used, from heart disease or diabetes to dementia or kidney failure. Patients may be unable or unwilling to come into hospital for tests and treatment, for practical or financial reasons, and they may stop taking their medication if they decide the side effects are too much to bear.

Another big problem is that these drugs are all mostly the same. Instead of giving us a toolkit equipped with all sorts of different widgets and gadgets that we can use on cancer cells, the pharmaceutical industry has presented us with a bag filled entirely with spanners with maybe a wrench or two thrown in there for good measure. Most of the therapies currently on the market are aimed at a relatively small repertoire of targets, mainly kinases and similar signaling molecules. This is partly a technical issue: it's relatively easy to find drugs that block overactive kinases as they have handy little biological pockets into which a drug can slip, much in the same way that a key fits into a lock. Many other products of mutated cancer genes are much harder to target and are often referred to as "undruggable."

To make things worse, the fact that such large financial returns are on offer for such slim survival gains has done much to encourage a culture of "Me Too."* Once one company develops a drug that successfully hits a particular target, others will race to make their own version that can get approved by being just a tiny bit better. This makes a certain amount of financial sense, as developing brand-new drugs is uncharted territory while developing a "Me Too" treatment relies on a molecular map that's already been drawn. Also, pharmaceutical companies are competitors and don't always play nicely

* This shouldn't be confused with the #MeToo campaign against sexual harassment, although this one is damaging and terrible in its own way.

together. Another reason to rush to get their own version of the latest hot drug is so they can make up their own combinations, rather than having to team up with anyone else.

Many new drugs are now approved by the slimmest of margins over their rivals, with a benefit of mere weeks often enough to gain the regulatory nod. Tarceva (erlotinib) was given the go-ahead for treating pancreatic cancer based on a study that showed just ten days' increase in survival. There's also some statistical wiggle room here. If companies test enough drugs, then a few of them will get over the line and appear to show a positive benefit purely by chance.

A commonly used measure for testing whether the results of a clinical trial are genuine and not just a random blip or fluke is something known as a "p-value of 0.05." Put simply, this means that if you repeated the same test twenty times, you'd expect to get the same result nineteen times and a different result once. This works for either positive or negative results: if you took twenty flavors of jelly bean and gave them to twenty groups of cancer patients, you would expect one flavor to coincide with a benefit in survival just through random chance.

Cancer drugs aren't jelly beans—they do contain biologically active chemicals that work in the lab and in animal tests. But given the number of drugs that are developed and tested every year, it's likely that some are sneaking under the statistical wire when they don't actually increase survival in human trials. And if the stats don't go in a company's favor, then marketing might still come to the rescue. I've spotted the weaselly small print "not statistically significant but clinically meaningful" accompanying the less-than-fabulous results of a trial of one of the latest wonder drugs.

There's yet another problem with precision oncology that's inextricably linked to the entire concept. Thanks to DNA and molecular analysis, we're now dealing with ever smaller groups of people whose cancer carries particular actionable mutations rather than a massive bucket of patients labeled "bowel cancer" or "breast cancer." And the billion-dollar economics of drug companies start to fall down when potential markets shrink to thousands of people or fewer—they'd much rather exploit an avalanche of identical patients than deal with a million special snowflakes. It's already hard enough to persuade drug companies to develop therapies for rare cancers, such as childhood tumors, so will they bother trying to hit targets that are just as scarce?

The never-ending churn of new expensive drugs and minimal increases in survival continues because there's no incentive to do anything differently. Patients and the public want new drugs, because they're sick of watching their loved ones die. Charities, companies, and academic or governmental organizations have invested eye-watering sums of money in the research that underpins these novel therapies. Regulatory agencies pride themselves on how many new drugs they can approve and how quickly they can get them through. And pharmaceutical companies stand to make an average profit of at least a billion dollars if they can successfully get a drug onto the market. We've fallen victim to the oncogene-pharmaceutical industrial complex, and it's financially unsustainable for all but the very wealthiest and well-insured.

Before you start thinking that I'm some kind of tin-foil-hat-wearing nutter posting conspiracy memes on the Internet, I certainly don't believe that this is the result of shady evildoers "hiding the cure." Cancer researchers and pharmaceutical industry staff are humans, too. All of us have lost people we love to this horrible disease, from family and friends to treasured colleagues. I've personally received distressing hate mail accusing me of being a "Big Pharma shill who wants people to die" even as friends, family, and co-workers were going through cancer treatment.

The pharmaceutical industry is still the best way we have of doing the large-scale research and manufacturing that's necessary to bring new drugs to market. Commercial organizations also bear the brunt of the substantial costs required to bring novel therapies through the lengthy clinical trials and regulatory processes. But I do think that many companies are guilty of all looking in the same place and thinking in the same way. And if they can reap a billion-dollar reward for getting a drug on the market that improves survival by a matter of months, there's no incentive to do any better. It's like promising to buy your child a car if they get a D grade in their exams: Why should they bother to work for an A?

We've made significant progress in cancer survival over the past couple of decades, particularly in wealthier countries. In the US, the proportion of people surviving cancer for five years has jumped from around half in the 1970s to two thirds today (although this varies widely by tumor type) and by ten years out that figure is lower again. Depending on your outlook, the glass is half full or half empty. There have been a few genuine game changers: cervical cancer screening; Glivec for acute myeloid leukemia; cisplatin for testicular cancer; major progress in childhood cancers; and immunotherapy for the small proportion of people in which it works. But genuine progress in advanced metastatic cancer is still grindingly slow. Life is short and precious, and there are far too many people who just don't have that kind of time. And because resistance is inevitable once a cancer has got to a certain size, I personally don't think that pinning our hopes on an ever-growing arsenal of magic bullets will bring the cures that we so desperately seek.

We do know that having certain mutations in your cancer means that you're likely to survive longer or shorter following your diagnosis. If you have a particular driver pushing your cancer cells to proliferate, then we can say that's a bad 'un or that's a good 'un. But right now we don't really have very good data to say that giving a specific drug based on that target will actually make a difference to your chances of survival, whether that's measured in months or years. We can't turn a bad cancer into a good one and if we keep doing what we've always been doing—finding cancer drivers and discovering drugs that block them—then we're going to get the results we've always got, just with ever more niche and expensive therapies. Survival rates might creep up, sure, but it's not going to be transformative.

As Vinay Prasad pointed out in a commentary in the journal *Nature* a couple of years ago:

> Precision oncology is inspirational. What doctor or patient would not want to harness genetics to tailor a therapy to an individual? But traveling back in a time machine is also inspirational. Who would not want to wind back the clock to remove their cancer before it spreads? In both cases, however, as of 2016, the proposal is neither feasible, cost-effective nor assured of future success. Yet in only one of these cases does the rhetoric so far outpace the reality that we risk fooling even ourselves.

We need to do better. We *have* to do better.

## OPENING THE COCKTAIL CABINET

If you speak to most oncologists and researchers about the emergence of resistance to targeted therapies, their replies tend to boil down to one word: cocktails. Combinations of different chemotherapies have been used in earnest since the 1950s, when pioneering New England pediatrician Sidney Farber began testing cocktails of drugs in his young patients with leukemia—at that time a universally fatal disease. Farber's success in turning flickers of hope into full-blown remission got everyone very excited by the idea that all cancers might one day be cured by the right drug combination.

Many conventional chemotherapies are given as

combinations of two or more drugs, which hit different mechanisms inside cells. For example, ABVD, a commonly used combination therapy for Hodgkin's lymphoma, is a combination of four different drugs: Adriamycin, Bleomycin, Vinblastine, and Dacarbazine. The first drug interferes with the molecular machines that unwind DNA for copying, the second causes DNA breaks, the third paralyzes the internal cellular scaffolding involved in cell division, while the last glues strands of DNA together so they can't be separated. But the combined side effects of all four together can be savage, not just during treatment but also causing heart damage, infertility, and even secondary cancers down the road.

Although chemo cocktails have been pretty successful in improving long-term survival for leukemia, lymphoma, and some solid tumors (notably testicular cancer), the hope that every cancer would have its perfect combination has turned out not to be true. Some cancers seem to be remarkably resistant to chemo by pumping the drugs straight back out again and there's always the capacity for resistance to evolve in other ways.

That idea is re-emerging in the Brave New World of precision oncology: sequence a tumor, find the drivers, then target as many of them as possible with a cocktail of therapies that are impossible to evolve around. It's a theory that owes a lot to the development of combination therapy for HIV in the mid-1990s (which, in turn, drew on Farber's earlier work on chemo combinations).

Drug companies had been developing and testing medications designed to target the molecules made by the virus for a few years, but they had limited effect when given individually—the virus just evolved resistance and the infection came roaring back. It wasn't until immunologist David Ho and mathematician Alan Perelson built a mathematical equation showing that the odds of a virus simultaneously evolving resistance to three different drugs were 10 million to one. This insight was truly life-changing and the resulting triple combination therapy, known as HAART, has led to a near normal lifespan for HIV positive people in countries where treatment is available and affordable.

A cocktail of targeted cancer therapies makes sense purely from a mathematical point of view, too: if different drugs hit different pathways in a cancer cell and require the evolution of different mechanisms of resistance, it's far less likely that a cancer will evolve resistance to both at the same time. For example, if a cell that's resistant to any single targeted drug is present at a relatively low frequency, say one in 100,000 cells, then given that even a small tumor typically contains more than 100 million cells, we can expect there to be at least 1,000 resistant cells in there. The treatment might kill off the 99.999 percent of the cancer, but that small pocket of resistance is enough for it to grow back again, as is sadly being abundantly confirmed in hospitals around the world every single day.

Combining two drugs raises the bar a lot higher: if each therapy kills susceptible tumor cells in different ways, then the chances of a cell having both mutations that render it resistant is effectively one in 10 billion. That's still within the bounds of possibility, especially for very large and rapidly evolving cancers, and trials of double-drug combos haven't had the success that many researchers had hoped for. But extending to three or four drugs makes it vanishingly unlikely that a cell with the full set of resistance mechanisms will be in there. The key thing is having a variety of drugs that work on different pathways driving cancer cells to proliferate. And that's something we simply don't have, thanks to the "Me Too" culture within the pharmaceutical industry.

We're currently at the stage where oncologists only have the ingredients to shake up a limited repertoire of recipes. If my drinks cabinet contains ten different types of vodka, three brands of vermouth, peach schnapps, and a couple of fruit juices I can rustle up a decent martini, Sex on the Beach or a Screwdriver. But I'm never going to be able to present you with an Old Fashioned or a Margarita. Our current collection of targeted therapies for cancer has a similarly limited palette.

Professor Bissan Al-Lazikani and her team at the ICR in London are using big data and machine learning in the hope of expanding the contents of the chemo cocktail cabinet. They started by picking out 470 cancer drivers—genes that are known to be mutated in cancer cells—homing in on around 120 encoding proteins that were suitable targets for drugs. Next, they traced all the interactions between different molecules inside cancer cells, figuring out which proteins "talk" to each other, and whether or not they switch another on or off and so on.

Eventually, they'd mapped out the whole social network—a sprawling web centered on a few key interconnected "hubs," similar to maps of air routes or Internet connections. Each of those hubs represents a cluster of related genes containing potentially druggable targets. But the targets that are hit by our current arsenal of cancer therapies are all clustered around a couple of hubs in one small corner of the network. You can hit that as hard as you like, but the rest of the network is still intact and can re-route around the problem, so the cancer cells will develop resistance and keep on growing.

Al-Lazikani and her colleagues have been using this map to tackle cancer in a smarter way, finding drugs that'll hit enough different targets to bring the network to its knees. It's an approach that's since been tested at the ICR using nearly fifty different types of bowel cancer cells grown in the lab. A combination of two targeted therapies that each hit a different signaling hub was enough to suppress cell growth for a while, but all the cells eventually evolved resistance to every drug pair. But nothing could survive the addition of an additional drug designed to block one of the "survival" molecules that prevents cells from dying when they're damaged. The third target is a completely separate part of the network, making it virtually impossible for the cells to evolve their way around this triple threat.

It's an important result, but there's a long way between showing that something works on cancer cells growing in plastic dishes in a laboratory incubator and having a safe, effective therapy that works for patients. To make matters worse, most drugs that start life in academic or industry labs will never even get near a patient. On average, more than nine out of ten new therapies will be canned at some point during the arduous journey from bench to bedside (perhaps having been labeled a "miracle" or "game changer" at some point along the way), with around half failing to show any benefit in clinical trials.

It's hardly surprising, given the high costs of research, that companies will be desperate to get their new wonder drug over the line of regulatory approval if there's even the slightest whisper of efficacy. But I find it absolutely shocking that such a large number of drugs get all the way through to being tested in humans—who volunteer their time and bodies to take part in trials—when they either don't work or barely make a difference.

Frustratingly, this well-trodden path of drug development from cells to animals (usually mice) to humans isn't actually leading us toward better treatments. We're great at finding new drugs that can cure cancer in mice, but very few of them work as expected once they get into the complex habitat of a human tumor. One solution is to use organoids: three-dimensional "mini-tumors" grown in the lab from patient samples, which behave in a more lifelike way than flat dishes of cancer cells or tumors transplanted into mice. Researchers are busy building huge banks of these organoids from all kinds of cancers, using them to screen for new drugs and combinations that might overcome the problem of resistance.

Another exciting new technology is "organs-on-a-chip." These are small glass slides riven with tiny microfluidic tunnels mimicking the plumbing inside a real organ, along with various types of human cells or molecules from that tissue, some cancer cells and all the nutrients they need to survive. It's still a synthetic system, but it's much more amenable to manipulations and measurements than a mouse (with the bonus of reducing the number of animals used in research) and provides a platform for large-scale high-throughput testing of new treatments. Add in the potential of three-dimensional printing with chemicals and cells to generate tissue or even whole organs in the lab, and the possibilities open up even further.

Al-Lazikani is hopeful that she might one day get to a point where she can build a chip that accurately recapitulates the organs or even the whole body of an individual patient, recreating the levels of molecules in their blood and adding in other medication they're taking, such as statins for high cholesterol. Then simply add some of their cancer cells and see which combinations of drugs are most effective and least likely to cause side effects.

There are other opportunities for expanding the oncology cocktail cabinet. In 2019, researchers published the first results from an ambitious study using precision molecular scissors (CRISPR) to snip out every single gene in thirty different types of cancer cells. They found more than 6,000 cases where knocking out just one gene was enough to kill the cells, which adds up to 6,000 potential new drug targets.

Many of these genes aren't appropriate targets for drug development, either because they're essential for healthy cells or because the proteins they encode don't have the kinds of molecular nooks and crannies that make good pockets for drugs. But even after discounting these tricky targets, the team still found 600 promising leads, most of which are outside the usual "Me Too" molecular repertoire.

Using technologies like CRISPR, organoids, and microfluidic chips to identify new drug targets and combinations that specifically target the molecular faults in tumors is the great hope for precision medicine. But while the idea of mixing up the perfect chemo cocktail based on the genetic makeup of each individual cancer is very attractive, there's only so much chemistry a human body can take.

HIV combination therapy works because the molecules made by the virus that enable it to replicate and spread are significantly different from those in human cells. This means that virus-specific drugs can be tolerated in relatively high doses without causing too much collateral damage—something that drug developers refer to as a "wide therapeutic window." Because cancers start from our own cells, there's a much greater chance that targeting the mutated molecules in a tumor will also have an impact on healthy tissue. Mix enough different drugs together and that therapeutic window slams shut.

The final big hurdle is regulatory rather than scientific. Right now, most new cancer drugs first have to prove their effectiveness as a sole agent in clinical trials in order to be approved. But this quickly starts to look nonsensical given how fast resistance can emerge to one or even two therapies and it's harder still to see a route for cocktails of multiple new drugs that are designed to work synergistically rather than separately. Even so, with the ticking of time and the mutational shuffling going on with any population of cancer cells, plus the selective pressure of the treatment, there's still a chance that a few straggling survivors will emerge.

So how do we go beyond whack-a-mole? Maybe the solution lies not in trying to target specific mutations and the inevitable emergence of resistant cells but in stepping back and looking at the whole mutational landscape within a cancer for clues as to how best to tackle it.

## MEET JOSH

When it comes to inherited mutations, Josh Barnfather got a bad one. He has a condition called *Xeroderma pigmentosum* (XP), caused by a fault in a gene encoding part of the repair kit in his cells that fixes damage from ultraviolet light. It makes his skin acutely sensitive to sunlight, so he can't go outside without sunscreen, long sleeves, and hefty black protective headgear that that looks a bit like a cross between a legionnaire's hat and a beekeeper's helmet. Despite these extreme precautions, his face is mottled with brown patches where UV has wrought its damaging effects and he's had more skin cancers than he can remember—at least ten in his three decades on the planet, all of which have been successfully removed with surgery.

Then in early 2017 he was diagnosed with something new. Something terrible. A type of cancer called angiosarcoma was growing just above his left eye and it wasn't going to go quietly. A pale patch of transplanted skin bulges over his eyebrow where the tumor was removed, but it was too late. A few months afterwards, he noticed swelling in his jaw as the cancer spread into his lymph nodes. Chemotherapy, radiotherapy, and a targeted therapy held it back, but by November 2018 the cancer had spread into his lungs, liver, and the lining around his heart. He was struggling to breathe and his doctors thought he'd be lucky to make it to Christmas.

He still hasn't met the woman who saved his life.

## THE CIRCOS IS COMING TO TOWN

Clinical geneticist Serena Nik-Zainal came into cancer research not from any particular passion for finding a cure but from an interest in technology. She became fascinated by Next Generation Sequencing—the revolutionary DNA-reading technique that finally opened up the possibility of analyzing tens, hundreds, or even thousands of whole cancer genomes at high speed and low cost.

By 2009, when she enrolled as a graduate student at the Wellcome Sanger Institute in Cambridge, researchers had become fixated on categorising cancers according to the genetic "shopping list" of mutations within them.* There was no longer such as thing as "breast cancer"—instead, there are at least ten specific tumor types, each defined by a particular set of genetic changes. There are at least four sorts of bowel cancer, each of which with its own characteristics and responding best to a different treatment strategy.

* "A very good way of making a scientific career but not a great way of treating cancer," as Nik-Zainal bluntly put it to me.

But as more data poured in from hundreds and then thousands of samples, it became increasingly clear that no two cancers are the same. Even cancers that start in the same tissue and look like they ought to fall into the same subtype can have completely different sets of driver mutations. And as researchers started to uncover the extent of the genetic patchwork of unique clones inside every tumor, providing fuel for the evolution of resistance, the picture only got even more complicated.

Now heading up her own research team at the University of Cambridge's Department of Medical Genetics, Nik-Zainal decided to take a step back from the gene-centric view of cancer. Instead of focusing on the presence or absence of specific mutations within an individual tumor, she's looking at the broader landscape of genetic damage, searching for common patterns between cancers that might explain the underlying biological processes driving the disease and reveal the most effective way to treat it.

When she and her team started looking at whole genomes from many cancers, they found that there were consistent patterns of damage that popped out, which started to make more sense than just trying to generate a shopping list of individual driver mutations. She pulls up a presentation on her computer screen showing the whole genome of a breast tumor taken from a 47-year-old woman, depicted in a type of diagram known as a Circos plot. This is a large circle with different-colored broken rings running around it, criss-crossed through with a number of straight and looping fine lines. It looks a bit like the kind of intricate mandala designs I used to draw with my Spirograph as a child.

What we're looking at is all the mutations in the whole tumor sample piled on top of each other, with no attempt to separate them out into individual clones *à la* TRACERx (page 173). The circle itself is a representation of all twenty-three human chromosomes, laid end to end. A ring of red dots picks out all the single-letter DNA "typos." Inside that is a ring denoting information about small insertions and deletions—green is where a piece of DNA has been gained, pink is a loss. Thin black lines criss-cross the center of the circle, tracing where whole sections of chromosomes have been swapped.

This particular breast cancer is nothing special, in genetic terms. Its genome has around 2,000 mutations, with a couple of key faults in known driver genes. The patient responded well to hormone therapy and is in complete remission. Nik-Zainal flicks to the next slide. It's another breast cancer that looks almost identical at first glance, but didn't respond to hormone therapy. Looking closer, even my untrained eye can spot the difference immediately. This plot is much more messed up, with many lines forming a spider's web across the center and much more intense bands of red and green. Curiously, these two tumors don't have a single genetic fault in common, despite originating in the same tissue and behaving in the same way.

"There are eleven thousand mutations, including four thousand deletions," she says, running a finger around the almost solid pink ring inside the circle. It's a pattern that she instantly recognizes as belonging to a tumor that has a mutation in a gene called MLH1, which means that it can't fix a particular type of damage to DNA through a process called mismatch repair. But while this mutation is relatively common in bowel tumors, it's not the kind of thing you'd go looking for in a breast cancer. Instead, this patient would get the standard treatment for hormone-positive disease, even though it doesn't have a hope of working.

This time, Nik-Zainal and her team could find the driver mutation in MLH1. But of all the samples she's analyzed that have this typical signature, only about half have a detectable fault in MLH1 or another relevant gene from the same cellular repair toolkit. The same turns out to be true for tumors with faults in either of the classic "breast cancer genes," *BRCA1* or *BRCA2*. These genes encode components of a different type of repair mechanism that patches up breaks where the DNA double helix is completely snapped in two (double-stranded breaks), and mutations in either of them leave telltale fingerprints in the genome. Confusingly, she was finding many tumor samples that clearly had this type of *BRCA* damage pattern but didn't seem to have any kind of underlying mutation in either gene.

In search of an explanation, she looked at breast cancers from more than 500 patients and found that more than a hundred of them had the typical signature of "*BRCA*-ness." Twenty-two of them were people with a known inherited fault in one of the *BRCA* genes, while thirty-three turned out to have a previously unknown hereditary mutation. Then there were twenty-two women who'd picked up a new *BRCA* mutation some time during development in the uterus. And then there was a final third with a *BRCA*-like pattern but no obvious underlying genetic cause. Clearly, Nik-Zainal and her team are still missing something important.

Discovering that a tumor has an underlying mutational pattern is important because it has a significant impact on treatment. In the case of *BRCA*-related cancers, they're particularly sensitive to drugs called PARP inhibitors, which block an unrelated "backup" DNA repair pathway. Left without any DNA repair mechanisms to call on, the cancer cells quickly become catastrophically damaged and die (a treatment approach known as synthetic lethality). These therapies are currently only available for women with certain types of cancer who have a known *BRCA* mutation. However, Nik-Zainal's results show that many, many more could be eligible.

"I just want to get this to the clinic," she says forcefully. "If one in five breast cancers are actually sensitive to PARP inhibitors then holy shit, we're missing out on a lot of people."

The moral of this story is that we should slow down in the relentless pursuit of driver mutations, as we may never find them. In many cases, alterations in the DNA sequence of a particular gene may not even be the thing that's responsible for changing its activity. There are all kinds of molecular flags and tags that are added or removed from the proteins that package DNA, known as epigenetic modifications, which affect how genes are switched on or off and are known to be messed up in cancer cells. Rearrangements in a chromosome can also affect how DNA is organized inside the nucleus of the cell, perhaps bringing an inactive gene next to a strong activator or dragging it into a "busy zone" where it's more likely to be switched on.

We're never going to find these kinds of influences by only looking for changes to the DNA sequence within a gene itself. Rather, we should look more closely at the overall patterns and processes at work and try to target them instead. It's a bit like knowing the modus operandi of a criminal gang but not caring about the exact identity of each individual mobster, because you're really only interested in nailing down the kingpin. If you wake up to find a horse's head in your bed, you don't need to know exactly which rogue put it there, but you certainly know which Family they came from.

Serena Nik-Zainal is working to make this kind of analysis a routine part of cancer diagnosis, in the same way that patients get routine blood tests and CT scans. She and her team can already draw up a mutational mandala from a tumor sample within twenty-four hours—far quicker than the current sequencing techniques used to look for actionable mutations. Doctors are busy, and there's a baffling array of tests and diagnostics already on the market, so this stuff needs to be simple and obvious. Starting with a way of identifying "*BRCA*-ness," she and her team are developing software that will sift through the complex landscape of mutations in a tumor sample and turn it into a simple output that can help doctors to decide how best to treat someone's cancer.

It was through this interest in matching the patterns of damage in cancer cells to the right therapy that Nik-Zainal started working with Dr Hiva Fassihi, who runs the UK's national service for people with XP—the condition affecting Josh Barnfather. And that's how she ended up with a sample of DNA from his angiosarcoma, taken during his initial surgery in 2017.

Because Josh's cells can't repair genetic damage caused by UV light, their DNA was shot through with the signatures of damage caused by sunlight. She shows me the Circos plot of his tumor genome, ringed by a thick circle of bright red from all the UV mutations. That wasn't surprising. But what was weird was the discovery that at a genetic level, it sure didn't look like an angiosarcoma, whatever his diagnosis said. Out of all the 800,000 changes in the tumor genome, there were none in what would typically be considered to be "angiosarcoma genes."

Then she looked closer. Hidden among all the mess of mutations from ultraviolet light was a different signature of damage. It wasn't very prominent and was only present in a small proportion of the cells in the tumor sample, but it was definitely there. One of the UV-induced mutations had hit a gene encoding a molecular machine called DNA polymerase epsilon (*POLE*), which normally copies DNA with extremely high accuracy as a cell gets ready to divide. Faults in *POLE* prevent cells from "proofreading" their freshly copied DNA, peppering the genome with thousands of spelling errors—a phenomenon known as ultra-hypermutation. At first glance this might seem to be piling more mistakes upon an already ruined genome. But Nik-Zainal's discovery had a profound implication: if a cancer has a mutation in *POLE*, it's likely to respond to immune checkpoint inhibitors.

These drugs—usually known just as "immunotherapy"—are the hottest thing in cancer treatment since the discovery of Glivec. They work by either overriding the molecular "off switch" that normally prevents immune cells from attacking tumors, or by interfering with the "secret handshake" that cancer cells use to persuade immune cells that they pose no threat.*

Unlike tests for targeted therapies, which rely on the presence or absence of a particular mutation, figuring out the underlying biological markers that determine whether someone's cancer is likely to respond to immunotherapy has been a lot trickier. There's growing evidence that checkpoint inhibitors tend to work best on cancers that have a very high degree of heterogeneity, but the presence of a mutated *POLE* gene—or even just the fingerprints of its handiwork in the genome—is one of the few positive indicators that we currently have. To Nik-Zainal, it was obvious—immunotherapy was Josh's only chance.

Before agreeing to prescribe him immunotherapy, Josh's medical team sent his original primary tumor sample for further laboratory testing to confirm it was suitable. The initial test came back negative, but Nik-Zainal wasn't giving up. Her analysis showed that the *POLE* mutation was only present in a subset of Josh's cancer cells, so would only be likely to show up in one of his secondary tumors rather than a small chunk of the primary cancer. She plagued the lab with

*There's a lot more about the science underpinning the development of checkpoint inhibitors and immunotherapy more generally in Daniel Davis' book, *The Beautiful Cure* (Bodley Head, 2018).

so many phone calls begging them to test some more samples that eventually they relented. Sure enough, another test on the primary tumor and a lump in one of his lymph nodes came back positive.

Even that wasn't enough to unlock access to immunotherapy. At the time of Nik-Zainal's analysis, the UK NHS would only provide these highly expensive drugs for a limited number of patients with lung cancer and melanoma, but not for patients with *POLE*-mutated tumors. Somehow, Josh needed to find the £60,000 ($75,000) required for the first three doses—an impossibly tall order for a PhD student without family wealth or hefty savings. There was one option left: crowdfunding.

Within a few months he had raised tens of thousands of pounds, not just from friends and family in his hometown of Hull, East Yorkshire, but from everywhere. There were raffles and street collections, pub quizzes, and Christmas sweater days. It was enough to get started with his first course of immunotherapy in early December of 2018.

When I Skyped Josh the following January, he was waiting for the results of his first scan since starting immunotherapy—a treatment that his oncologist believes came just in time to save his life. He was nervous but hopeful, tipping his shaved scalp toward the camera to show me the gentle undulation on his crown that had previously been an angry lump. By February, his update on the crowdfunding platform confirmed the good news he'd been wishing for and the treatment was continuing to work well by the end of 2019 with very few side effects. I'm keeping my fingers crossed for him.

## TIME TO EVOLVE

So, where are we now when it comes to treating advanced metastatic cancer? There's a huge amount of time, money, and effort going into precision oncology, despite the obvious and well-documented issues with the evolution of resistance to almost any drug we can currently throw at the disease. It's still proving hard to wean drug companies off their addiction to "Me Too" compounds with ever smaller survival margins and get them interested in exploring new targets and cocktails, even if the combined side effects of multidrug combinations don't prove to be too toxic to handle.

Immunotherapy is incredibly exciting, but it doesn't work for everyone. We still need to work out how best to select the patients whose cancers will respond to these drugs, like Josh, and find alternative ways of awakening the immune system in people who don't. And cancer cells can still evolve ways of evading the immune predators roaming the tumor habitat, going into stealth mode by switching off production of the proteins that make them conspicuous to immune cells.

If we're going to make a real difference to cancer survival in the future we need to think properly about the problems—and the potential—in evolution. We need to be smart. As smart as Charles Darwin himself.

# 10

# GAME OF CLONES

If you've ever watched *Avengers: Infinity War* you'll remember this scene. A ragtag band of superheroes are up against the ultimate baddie, Thanos, who's gathering a full set of infinity stones in order to wipe out half the universe. There's a bit near the end of the film where Doctor Strange is sitting in the rubble of a destroyed planet, glitching and twitching as he leaps forward in time to glimpse all the possible futures that await the group in their coming battle.

"How many did you see?" says Star Lord.

"Fourteen million, six hundred and five," Strange replies.

"How many did we win?" asks Tony Stark.

Pause. Cue dramatic music.

"One."

In many ways, the Avengers' situation feels a lot like the challenge of treating advanced metastatic cancer. There are so many mutations, so many cells, so many options to evade and escape, that any hope of a cure seems vanishingly improbable. But what if we could see into the future, like Doctor Strange? What if we could look at the messy, mutated patchwork within a tumor and the devastated habitat surrounding it and predict how it'll respond to any possible treatment, be that chemotherapy, radiotherapy, precision drugs, or immunotherapy? Instead of playing whack-a-mole, reacting and responding to the emergence of each new threat, we'd be in the driving seat, knowing that the next move we make is steering the evolution of the disease exactly where we want it to go.

## FROM MOLECULES TO MATH

On the fourth floor of a nondescript academic building at the Moffitt Cancer Center in Tampa, Florida, Robert Gatenby and his colleagues are testing out a new weapon in the war: math.

A clinical radiologist by training, with a deep-seated interest in mathematics, Gatenby has assembled a curious band of experimental biologists, mathematicians, data scientists, physicists, and clinicians. The challenge is finding a way to make them all talk to each other. Most of the floor is open-plan office space, designed to foster cross-disciplinary interactions. There's also a central "Collaboratorium"—a large room designed for informal, spontaneous discussions, lined with chalkboards at the insistence of the mathematical traditionalists in the team. There's one token whiteboard for those who don't wish to get their clothes dirty, along with a computer projector for non-mathematicians, who prefer PowerPoint presentations to scribbled equations.

Gatenby first realized that there must be a different way to approach cancer when he read an article about the diamondback moth—a voracious crop pest plaguing farmers for more than a century, which has now become resistant to virtually every pesticide on the market.* It was immediately obvious to him that what had happened with the moths was exactly what was happening in cancer—the evolution of resistance to therapy, followed by unstoppable spreading.

One of Gatenby's biggest peeves with current cancer therapy is the way that drugs are prescribed, usually referred to as the "maximum tolerable dose" (MTD). This is the largest amount that can be given to a patient before they suffer unacceptable side effects, in the hope of getting rid of as many cancer cells as possible in one massive chemical hit. The maximum dose for most treatments is figured out during early stage clinical trials by gradually increasing the dosage in successive groups of volunteers, stopping short of serious side effects. But, as we've seen, resistance is virtually inevitable, causing a lot of toxicity for a relatively modest survival benefit.

Looking to the example of diamondback moths, farmers have been dealing with this problem of pesticide resistance for decades through an approach known as Integrated Pest Management. Much like a tumor made of genetically unique cell clones, some of which will be resistant to therapy, a population of crop-destroying insects will be a mixture of genetically diverse critters. Some of them can be killed by bug spray while

* Gatenby admits that his attention was first drawn to the piece because the moths eat cabbage, which he hates.

others will shrug it off. Importantly, the gene variations that make insects resistant to pesticides tend to make them a bit worse at feeding and breeding ("less fit" in evolution-speak), so they're outcompeted by the pesticide-sensitive strains and stay as a small proportion of the population under normal circumstances.

If you hit this mixed gang of bugs with a massive dose of insecticide, you'll wipe out all the sensitive ones and leave a hardcore of resistant pests that are now free to munch and mate as much as they like. But if you just aim to knock the population down to a reasonable level, rather than aiming for total eradication, there will be enough of the sensitive bugs left to outcompete the resistant ones and keep them under control.

Rather than trying and failing to wipe out every last critter, farmers are learning to live with them. They regularly monitor the insect populations and put up with a certain level of destruction in their crops, only using pesticides when things get out of hand. Similar approaches are now being used to control weeds and other unwanted species—even the prolific diamondback—but the principle is the same: aiming not to eradicate but to control, reducing collateral damage by pumping fewer toxic chemicals into the environment. In turn, this cuts the chances of generating unstoppable superstrains that will cause much more trouble further down the road.

The parallels with cancer are obvious. Some tumors can be completely cured through surgery—the equivalent of chopping off a diseased branch from a tree, squishing a bug, or pulling up an isolated patch of weeds. As the surgeon's proverb goes, "Nothing heals like cold steel." Other cancers are a minor infestation, small and homogeneous enough to be eradicated with a quick spray of chemotherapy and mopped up by the predatory cells of the immune system (which could be compared with biological agents such as the tiny worms that are used to control slugs). But advanced metastatic disease is like the diamondback moth. It's everywhere and we're out of options to control it.

We have to stop ignoring the evolutionary processes happening right under our noses and not only accept them, but also start using them to our advantage. We've tried (and failed) to eradicate cancer by throwing the chemotherapy equivalent of Agent Orange at it, so Gatenby wondered if there was a way of taking the strategy of Integrated Pest Management and applying it to controlling the disease rather than aiming to cure it.

Rather than ignoring the possibility of resistance, he started from the assumption that a small pocket of drug-resistant but less fit cells will be present in a tumor from the very start, growing to fill the void left when sensitive cells are killed off. Instead of using the maximum tolerated dose of a drug to shrink a tumor as small as possible, the aim is to use a lower dose that reduces the cancer to a certain level, so that there are still enough sensitive cells around to keep the resistant ones in check. Once that population of sensitive cells gets too big, they're given another dose of the same drug, which knocks them back down again. It's an approach that Gatenby calls "adaptive therapy," harking back to the idea of playing evolution at its own game.

Adaptive therapy relies on the fact that while being resistant to treatment is an advantage for cancer cells in the presence of chemotherapy, it's a disadvantage under normal conditions as it costs a lot in biological terms to throw off the effects of drugs. For example, the molecular "pumps" that eject chemotherapy out of resistant cells can take up to a third of a cell's total energy budget, leaving little spare for proliferation. Some cancers can become "drug addicts," ending up hooked on the chemicals that are supposed to do them harm. Melanoma cells that have rewired their internal pathways to become resistant to a particular targeted therapy are unstable and prone to dying if the drug is taken away. Under normal conditions without treatment, these costs stack up and the resistant cells will grow more slowly. Gatenby uses the analogy of carrying a large, unwieldy umbrella around the place: it's incredibly useful if it's raining, but just gets in the way and slows you down the rest of the time.

There have been some previous attempts to use on-off, on-off cycles of treatment to improve survival from advanced cancer, but trials of this approach (known as metronomic chemotherapy) haven't been particularly successful. That's because the cycles of treatment didn't take into account the status of the underlying struggle between the resistant and sensitive cells. Simply dropping another dose of chemotherapy without knowing which population is currently winning the tussle is as unhelpful as dropping a bomb on a battlefield without knowing the locations of each set of enemy combatants. You might make a few lucky strikes, but do it enough times and at some point you'll make a disastrous error.

To understand the evolutionary dynamics at work, Gatenby and his team started with math. Using measurements from lab experiments, they figured out a set of equations based on how quickly the sensitive and resistant cells were growing and the impact of drug treatment on each. Then they used their calculations to create a virtual simulation of how the two populations would grow and shrink once treatment commenced and plotting out exactly when to give each dose of therapy. Satisfied that the approach should work to control tumor growth in the long term, they tested it out on mice that had been injected with ovarian cancer cells.

Once the tumors had grown to the size of a small pea, each animal was given a shot of carboplatin, a drug typically used to treat the disease in humans, with a few animals left untreated as a comparison. The treated mice were then split into two groups. Some of them got a standard dose of carboplatin every four days like clockwork, mimicking a metronomic schedule, while others went on an adaptive therapy regimen. Every three days, one of the researchers would carefully measure the size of each animal's tumor with a set of tiny calipers and adjust the dose of drug accordingly—a bit more if it was growing, a bit less if it was shrinking, according to Gatenby's calculations.

Even though this was a fairly scrappy experiment in a highly artificial system, the results were striking. Six months after the initial tumor transplant, the cancers in the animals receiving regular doses had more than quadrupled in volume and were actually slightly larger than the mice that hadn't been given any treatment at all. However, the tumors in the animals on the adaptive therapy were the same size they had been at the start. There were a few ups and downs along the way, but it worked.

Maybe, though, they had just struck lucky with this ovarian cancer model. The next challenge was to try adaptive therapy on a different cancer with a different treatment. This time, they looked at mice injected with two types of breast cancer cell lines—one from a typical hormone-sensitive tumor and the other from a so-called triple negative cancer, which is harder to treat—and compared standard versus adaptive dosing of a drug called paclitaxel.

Five years after their first ovarian cancer paper, the Moffitt had finally acquired a mouse-sized MRI machine, so Gatenby and his team were able to carry out more sophisticated measurements of tumor size in order to fine-tune the treatment. Again, the results were remarkable. The tumors in the animals receiving adaptive therapy stabilized and stayed small, using lower and lower doses of the drug over time. In a few cases, the cancer remained stable even when the treatment was stopped completely. It worked.

It was time to scale up. Although the animal experiments had been done with breast and ovarian cancer, Gatenby decided to run his first trial with prostate cancer patients. It was purely a practical decision: a vital component of adaptive therapy is the ability to monitor the change precisely in the amount of cancer cells in the body (tumor burden) on a regular basis, ideally with a non-invasive test. Putting patients through frequent CT or MRI scans is logistically difficult and expensive, and also exposes them to an unnecessary amount of X-ray radiation (in the case of CT scanning). But prostate cancer cells produce a chemical called PSA, which is measured with a simple blood test that can be done at a local doctor's surgery. It's not a perfect read-out, but it was good enough.

Gatenby teamed up with Jingsong Zhang, a prostate cancer specialist at the Moffitt, who had seen the results of the mouse studies and was keen to get involved in a clinical trial. The aim was to recruit a small number of men with advanced metastatic prostate cancer who'd tried every treatment except one—a pricey new drug called Zytiga (abiraterone), which shuts off the production of tumor-fueling testosterone. At this point in the journey, most men getting a standard schedule of the drug could expect an average of eighteen months or so before their tumors evolved resistance and started growing again.

The team spent hours poring over the right numbers to plug into their mathematical model, thinking about the populations of cells that were likely to be present in a typical prostate cancer and how they'd respond to the drug, as well as what might be feasible from a practical point of view. Although the calculations were complicated (Gatenby had to leave the equations out of the clinical trial approval application for fear of scaring the committee), the idea was fairly simple: at the beginning of the trial, all of the men's PSA levels get tested. Then they start on regular daily doses of abiraterone, with PSA tests every four weeks and a CT and bone scan every three months.

Once their PSA drops to half the starting value, each

patient stops taking the drug. And then they wait. At some point—weeks, months, it depends on each man's individual cancer—their PSA starts to go back up as the tumor begins to grow again. Once it's back to the same size it was at the start, they go back on the abiraterone and the cycle repeats. Or at least that's the idea.

Following the first patient through the study was a nerve-wracking experience. Even though Gatenby was fairly sure it was going to work, he admits to feeling nervous when the man stopped taking abiraterone for the first time and his PSA began to climb.

"I was thinking, *What if we're wrong? What if it doesn't work and it just keeps going and we give the therapy again and it doesn't work?*" he says, grimacing at the memory. "Then of course you have to watch it come back up again on the next cycle. I have more confidence now, but I always worry—I want to do the right thing for people and I don't want to hurt anybody."

This desire to first do no harm meant that the trial took a lot longer to recruit patients than anyone expected. The team had agreed that if the first three patients didn't show the expected cycles of tumor shrinkage and growth, then they would stop the study. But while their mathematical models predicted that a cycle could be anything from three months to a year and a half, they hadn't anticipated quite how variable each patient would be. One of the very first patients recruited onto the trial had an incredibly long cycle, taking the best part of a year for his cancer to grow back to the level where treatment could commence again. It was good news for the patient, of course, but incredibly frustrating for Gatenby and his colleagues.

For the patients on the trial, this was the hardest thing to get their heads around. Why would you stop a treatment that's working before it's fully done its job and deliberately let the cancer come back? It just didn't seem to make sense. For two of them, it was more than they could take. Once their PSA levels had dropped to 50 percent they insisted on carrying on with the treatment, driving them down as low as possible. In both cases, they relapsed very quickly and died.

To some of the other participants, the idea behind adaptive therapy makes perfect sense. Gatenby tells me about his most powerful patient advocate—Robert Butler, a British oil industry engineer who had moved to Tampa to enjoy his retirement and was diagnosed with prostate cancer in 2007. Hormone therapy and radiotherapy had failed to keep the disease at bay, and he ended up on the adaptive therapy study as his last hope.

At the time of my visit to the Moffitt in May 2018, Robert was on his tenth cycle of abiraterone and holding the record for survival on the trial. For someone with a mechanical mind, he could see the obvious parallels between the judicious application of a drug to tweak and balance the population of cancer cells in the body and a device like a thermostat, which only kicks in when things get a bit too chilly. The temperature in the room might fluctuate as the heating cycles on and off, but the overall outcome is fairly stable.

For the men who stuck it out and rode the abiraterone roller coaster, the results have been impressive. The preliminary results were published at the end of 2017 and showed that ten out of eleven patients were completely stable, with their tumors undergoing well-behaved cycles of growing and shrinking for an average of twenty-seven months—nearly a full year longer than would have been predicted had they taken the standard schedule. Only one patient's cancer failed to respond as predicted and got worse. Importantly, the men ended up taking an average of around half the usual total dose of abiraterone—some of them only taking the drug for one month per year—and there were very few side effects, none of them serious.

Although adaptive therapy aims to stabilize cancer by controlling the population of drug-resistant cells within a patient's body, they're still in there somewhere and they're still proliferating (albeit relatively slowly). Gatenby's mathematical models suggest that it should be possible to get through up to twenty cycles of treatment before the population of resistant cells takes over, but it will probably still happen at some point.

By February 2019, when I saw Gatenby's colleague Joel Brown present the latest results at a meeting in Paris, France, a total of sixteen men had been recruited onto the trial. On average, they'd lasted twice the length of time before relapse than would be expected, with one man making it past four years. Unfortunately, all of them relapsed in the end, as the drug-resistant cells eventually took over, but most were still alive and trying other last-ditch therapy options. And they were all doing far better than a similar group of men taking abiraterone on a conventional schedule. By any measure, the study was a success. *It worked.*

In theory, adaptive therapy should be applicable to any kind of cancer or treatment, as long as there's enough knowledge about the various types of resistant and sensitive cells in order to create a mathematical model, along with a way of regularly measuring tumor burden. This is a major stumbling block right now, and there's an urgent need to develop simple, cheap, and non-invasive ways of following the progression of cancer without having to rely on repeated scans.

The most promising idea in this area is the concept of "liquid biopsy"—analyzing the levels and genetic makeup of tumor DNA or cells shed from tumors into the bloodstream, or searching for more molecules like PSA that provide a reliable read-out of tumor burden. There's a lot of excitement around using liquid biopsies for monitoring how cancers change in response to treatment, as well as maybe selecting the most appropriate therapy or even diagnosing the disease in the first place. It's definitely one to watch in the future.

Moving away from treating cancer by hitting it hard with maximum doses toward a more nuanced evolutionary approach also raises some intriguing questions about the kinds of drugs that are best suited to long-term control. Ironically, because adaptive therapy aims to maintain the balance within a tumor, rather than destroying as many cancer cells as possible, it might work better with what would conventionally be considered to be "less good" drugs—therapies that don't have such a dramatic kill rate but have fewer side effects and lower long-term toxicity.

The shelves of pharmaceutical companies are stacked with many such molecules, rejected for not being effective enough in lab tests or clinical trials, which are ripe for reinvention in a subtler adaptive context. And there are plenty of existing drugs that are already approved and being used for other diseases—such as anti-parasitic agents and heart drugs—many of which have low toxicity and might be useful for treating cancer in an evolutionary context even if they don't "work" according to the conventional maximum-dose mode of thinking.* As an added bonus, the protective patents on many older treatments are likely to have expired. Normally, companies are granted exclusive rights to sell a particular drug for a certain period of time after it has been approved, enabling them to recoup the costs of development and compensate for the failures that never make it to market. Off-patent drugs are potentially cheaper to produce as generic medicines than newer therapies, which not only makes them attractive to healthcare providers but also expands access for countries with less money available for cancer care (although I would never underestimate the pharmaceutical industry's ability to turn a profit.)

The contrast between the Moffitt's prostate cancer trial results effectively showing a doubling of progression-free survival from metastatic prostate cancer and the mere months that are squeaked out of a typical novel targeted therapy couldn't be more stark. If this was a new kinase inhibitor, drug companies would be falling over themselves to get it

* The ReDO project (Repurposing Drugs in Oncology) is an international collaboration of researchers working to identify promising cancer treatments based on existing therapies for other diseases. Find out more about their work at redo-project.org

registered, but there's been very little interest in applying evolutionary theory to cancer treatment. Ideas for long-term control rather than cure get crowded out of the discourse by talk of "magic bullets," precision medicine and ever more expensive therapies. Ironically, adaptive therapy is exactly the kind of thing that conspiracy theorists often throw at the pharmaceutical industry—claiming they want to keep people sick rather than come up with cures—but it's taking a long time for the mainstream to sit up and take notice.

## ALL SWEAT, NO GAIN

The first time I heard about adaptive therapy, the hairs on the back of my neck stood up. It felt like something so new and fresh—a sea-change from the rising tide of targeted therapies with minuscule survival benefits. News of the Moffitt's prostate cancer study has spread and there are stories of oncologists in other American hospitals trying the same approach with their patients. Gatenby and his team are now embarking on a new prostate cancer study, testing adaptive therapy as a first line of treatment rather than a last resort. If they can keep the roller coaster going for years or even decades, then this starts to look like a viable way to turn advanced cancer from a short-term killer into a long-term chronic condition.

My personal enthusiasm isn't enough to convince doctors, regulators, or patients, however. It would be foolish to get overexcited about the results of one trial and a handful of animal experiments. The most important proof will be to show that it works over and over again, in more and more patients with all kinds of cancers.

Right now, there are more adaptive therapy trials planned across a range of tumor types, from more common diseases like metastatic breast cancer to deadly rare childhood cancers and brain tumors, where any gains in survival could make a real difference. Although resistance will probably creep in over time, as it has done with the men on the prostate trial, there may be the option to try a second or even third line of treatment after that. It's still whack-a-mole, but on a much slower timescale.

There are plenty more ideas where this came from, using evolutionary and mathematical principles to move from short-term failure to long-term cancer control or even total extinction. One of my favorites is *ersatzdroges*, or "decoy" drugs. Many tumors develop resistance to therapy by switching on banks of molecular pumps that shuttle chemotherapy drugs straight back out of the cancer cells before they can do any harm. These pumps use a huge amount of energy, but it's a worthwhile investment to stay alive. Gatenby's team discovered that dosing chemo-resistant cancer cells with an old and relatively non-toxic blood pressure drug, verapamil, was enough to keep the pumps spinning at full speed, wasting a huge amount of energy and leaving them with little fuel left in the tank for proliferation ("all sweat, no gain," as their paper describes it).

By contrast, chemo-sensitive cells have a competitive advantage and grow fast but can be easily knocked back with each round of treatment. Using *ersatzdroges* to hobble the growth of drug-resistant cells could be a valuable addition to adaptive therapy, ensuring that resistant cells don't grow faster than their sensitive neighbors in between cycles of treatment. There might be other ways to make drug resistance very costly in biological terms. For example, if resistant cells are particularly dependent on certain nutrients, such as particular amino acids (the building blocks of proteins), then selectively creating a starvation situation would make life harder for them and easier for sensitive cells to thrive.

Another revolutionary evolutionary idea is the double bind or "sucker's gambit." The idea here is to steer cancer cells down one evolutionary path before hitting them with a new threat against which they're now unable to defend themselves. Imagining this in the natural world, it would be a bit like a species of fast-breeding mouse that has two predators: snakes and hawks. Mice scuttling around in the open are liable to be picked off by birds, while those seeking shelter in the undergrowth are more likely to fall victim to snakes. As a population, the mice are now in an evolutionary double bind and unable to adapt to life in just one habitat, as neither is safe for them. For cancer, the equivalent would be using two therapies with mutually exclusive resistance mechanisms—if cells adapt to grow in the presence of one treatment it means they must be inescapably sensitive to the other and therefore unable to thrive in the presence of both.

Yet another concept is "benign boosters," proposed by Carlo Maley (the man with the radiation-resistant sponges we met in Chapter 1). These are drugs that encourage the growth of benign, non-invasive cells in a tumor, which outcompete and suppress any aggressively malignant clones with expansionist ideas. There are a few obvious rules that must be obeyed: the benign cells should be easy to control if they get out of hand (for example, with chemotherapy) and they must have a competitive advantage over aggressive cancer cells. Computer simulations showed that benign boosters could be effective at controlling advanced metastatic cancer or preventing relapse even when applied late in the game. This logic could also be applied to cancer prevention, boosting the growth of healthy cells to crowd out any cancerous clones, and Maley and his team are now working out how best to test this idea in real life.

So far they've carried out tests on cells growing in plastic dishes in the lab, finding that the chemical ascorbic acid helps normal esophagus cells to outcompete precancerous cells found in a condition known as Barrett's esophagus. Ascorbic acid, more commonly known as vitamin C, is found in fruit and veg, as well as dietary supplements. We have absolutely no idea whether swallowing supplements or swigging fruit juice could prevent esophageal cancer cells from gaining the upper hand in humans, and I would never suggest using a single lab-based paper about vitamin C as a recommendation for any medical intervention.* But this intriguing preliminary finding highlights the need for more systematic investigation of ways to balance the interactions and competition between

* Nobel laureate and biochemist Linus Pauling was particularly enthusiastic about the use of high-dose vitamin C to treat cancer, spawning legions of dubious claims on the Internet despite the lack of convincing clinical evidence.

cancer cells and their healthy counterparts, or playing off the clones within a tumor in favor of more benign ones.

## REBELLION EXTINCTION

I've talked about using evolutionary strategies to control cancer, transforming a short-term killer into a long-term condition. But what about using them to cure?

What we're really talking about here is an extinction event—the total eradication of a population from an environment. It's something that's happened many times throughout the history of this planet, whether by *force majeure* (climate change, stray asteroids, diseases, et cetera, et cetera) or at the hands of humankind. More than 99 percent of species that have ever lived on Earth have vanished—many of them more numerous and genetically diverse than the cells in a typical tumor—so there's much we can learn from extinctions in nature if we want to engineer an evolutionary endgame against cancer.

When we think about extinctions, one of the first events that come to mind is the asteroid strike that struck out the dinosaurs—a massive global catastrophe around 66 million years ago that wiped out three quarters of all the animals and plants on Earth. But while this did indeed wipe out all the giant dinosaurs, the smallest were able to survive and evolved into modern birds.*

* If this were a metaphor for cancer, the asteroid would be maximum-dose therapy and the survivors would be the resistant cells, eventually taking flight and crapping all over the place.

Most extinction events are much less grand, with individual species tending to go out with a whimper rather than a bang. The first step is some kind of crisis, such as intense predation or hunting, loss of habitat, or a change in the environment, which reduces the population to a small bunch of survivors that are ill-equipped to cope with any more sudden changes. Small populations have low genetic diversity, meaning that individuals tend to share many of the same gene variations, so they're less able to adapt and evolve their way out of trouble. There's also a greater chance of health problems and susceptibility to diseases owing to inbreeding, and the risk of a few unlucky events wiping out the entire population is high. At this point, life becomes extremely precarious and extinction is the most likely outcome.

In a paper written together with evolutionary biologist Joel Brown, Bob Gatenby tells the story of the heath hen as a demonstration of extinction in action. These charming wild fowl were widespread over the east coast of North America when the first European settlers arrived. Predictably, the lure of the large grouse-like birds proved to be irresistible and the colonists took to hunting and eating them with great enthusiasm. In fact, the original Thanksgiving turkey may have actually been a heath hen.

As the European settlements expanded, the hen's habitat shrank. By 1870 there were just fifty birds nestled on the island of Martha's Vineyard off the New England coast. The islanders rallied round to save their feathered friends and the population grew to around 2,000 over the next few decades. Then fate intervened. A fire ravaged the breeding grounds, followed by a few unusually harsh winters. Finally, an infectious disease arrived and polished off the remainder. The last heath hen, known to locals as "Booming Ben," died in 1932.

Colonial expansion and hunting acted as a "first strike" on the population, knocking it down to a perilously low level—an outcome that was depressingly predictable based on similar situations elsewhere. The hens were forced into a geographical and genetic bottleneck—a small breeding population stuck in one place—which put them in a risky situation. The subsequent smaller strikes were more random and harder to predict, but the species was already well on the road to extinction.

All over the world, conservationists are studying shrinking populations and habitats, developing mathematical and genetic models to figure out the risk of extinction and work out the best strategy for trying to save struggling species. And because cancers are also populations of cells living in the habitat of the body, the same models will apply. But rather than trying to save these cells and encourage them to thrive, we deliberately want to drive the numbers to zero.

As Gatenby and Brown point out in their paper, this model is already in use for treating children with acute lymphoblastic leukemia (ALL), although not usually viewed in such explicit evolutionary terms. Over the years, doctors have used trial and error to figure out a life-saving combination of treatments delivered on a specific schedule, transforming the disease from universally fatal to curable in nine out of ten cases. In doing so, they've hit on exactly the same strategy that doomed the heath hen. There's a big first strike—intense chemotherapy—that kills the bulk of the population of cancer cells, leaving behind a small band of survivors. Then there's a second hit with a drug that has a different method of action, which hits cells that were resistant to the first one, followed by a third and a fourth.

Instead of coming up with combinations through a lengthy process of iteration, Gatenby's mathematical models make it possible to start devising extinction strategies from scratch. When viewed as an evolutionary and ecological extinction problem—reducing a genetically diverse population to the point where it will collapse and can be wiped out—these ideas seem blindingly obvious. But they run completely counter to the way that patients are usually treated today. For example, when men with advanced prostate cancer are treated with a hormone-blocking drug like abiraterone, they'll usually be given a maximum dose over a long period of time. This covers not only the time taken for the tumor to shrink down but also as long as it takes for it to grow all the way back again, chock-full of resistant cells.* At this point, the doctor might suggest switching to a different chemotherapy and going round the loop again.

But if you're aiming for extinction, then why wait until a large population of tumor cells has grown back? Surely it would be better to kick 'em when they're down. The best time to use the second drug is actually when the population of cancer cells is smallest, just after the "first strike" of

* Adaptive therapy obviously gets around this problem of wholesale resistance, but it's only aiming for control rather than cure.

abiraterone has done its damage. Any surviving cells will be sickly and weak, owing to the effort of maintaining resistance against the therapy, and a second hit with cell-killing chemo is likely to finish them off. This idea appears acutely counter-intuitive—why switch therapies when the first one seems to be working?—but it's much more likely to drive the disease into the ground.

## A GAME OF LIFE AND DEATH

By combining evolutionary and mathematical knowledge to understand the problem of cancer, Gatenby and his colleagues are treating the disease as a game. Although using the word risks making a serious topic sound trivial or simplistic, game theory is a well-developed set of mathematical rules that can be applied to all kinds of interactions between individuals, whether that's people, animals, or cells. By understanding the rules by which cancer cells adapt and respond to therapy, we should be able to beat them at their own game.

Importantly, our team has a major advantage: sentience. Oncologists are (more or less) rational creatures. They make deliberate strategic decisions about which treatments to use and when, and know what they'll do next. Cancer is purely reactive, adapting in response to changes in the selective pressures in its environment—the presence of drugs, low oxygen, depleted nutrients, whatever—but is unable to predict what lies ahead. Giant dinosaurs were doing great until an asteroid strike completely changed the environment of the planet. If conditions suddenly change in a way that's incompatible with the characteristics that have evolved within cancer cells to ensure their survival, they may quickly find themselves on the road to extinction.

With this ace up our sleeve, it's time to play. Unlike a game such as "rock, paper, scissors," where both players make their move at the same time, the game played by doctor and disease is a Stackelberg game, first developed by German economist Heinrich von Stackelberg and played turn by turn. This gives the first player an advantage, as the follower has to make their move in response to this opening gambit, restricting their options.

An example might be something like tic-tac-toe, where there are a limited number of places to put an O once the first X is in place. As I know from beating my younger sisters frequently when we were children, if you go first against a less experienced player, you tend to win. The same should be true of a Stackelberg game played against cancer: if the doctor makes the first move and each individual patient's cancer is a newbie at the start of treatment, then we should be able to work out the rules that enable us to win every time. Unfortunately, this isn't what happens at the moment.

Right now, an oncologist may make the first move by picking a particular drug, but then they lose their advantage by continuing with the same treatment until resistance emerges. Once they switch to a different therapy it's too late: the cancer is in the driving seat and every further treatment option is a response to its next move. It's like starting a game of tic-tac-toe by drawing an X, waiting for your opponent to draw their O, then drawing the same X again in the same box. You've lost your first mover advantage and all you can do is follow. And you're probably going to lose.

As I look around Gatenby's office on the fourth floor of the Moffitt, I notice that the floor is covered with colored squares of carpet tiles, similar to a chessboard. It's also quite large.

"Can we play a Stackelberg game? Right here?" I ask enthusiastically.

We decide the rules. I'm the doctor, he's the cancer. It's a chase, where I win if he can no longer move. We square up, one tile between us. I give him a wicked grin and battle begins. To start with, I play the "conventional MTD therapy" gambit, continuing with the same treatment until it becomes obvious it's no longer working. I take a step forward. He takes a step back. I take another step forward. He steps to the side. He's evolved resistance. I take another step forward—after all, I'm still getting closer to him so why would I change my strategy now? His next step brings him level with my shoulder. I take another step forward, passing him completely. He's free and has won the game, while I'm quickly heading straight toward the wall. Even if I had switched tactics once he was level with me, I'd lost the moment he sidestepped.

We reset for another round. This time, I'm going to be smarter. I step forward. He steps back. I step forward again. He steps to his left. I follow his lead, stepping to my right. He moves sideways, back the way he came. I follow. We dance like this for a while with much giggling, shuffling from side to side but never making progress past each other. It's not ideal—and eventually one or both of us will probably get tired and give up—but it's effectively a draw.

We start again. Finally, I come up with an idea for how to win. I step forward as before. He steps to the side. I pull the trash can out from under his desk and put it in front of me, then step to the side as well. His sideways escape route is now blocked. He has to step backward. The chase is on, and I've soon backed him into the filing cabinet.

"I lose!" he laughs. "That was fun!"

I was able to win for three reasons. First, Gatenby's office is a finite space, and there were only so many places he could go. Secondly, I was watching his every move and changing my strategy, predicting where he was likely to go next. But because I had the first mover advantage, I was always in the driving seat and leading each round. And finally, I brought out my secret weapon: the trash can.

Cancer cells might have the entire human genome to play with, shuffled and shattered with a multitude of mutations, but their options aren't infinite. Although there may be many genetic routes by which cancer cells adapt to the stresses and strains placed upon them, they tend to converge on similar outcomes. Evolution may indeed be smarter than we are, but we have the benefit of science on our side. By studying the evolutionary responses in hundreds and then thousands of cancer patients, we should be able to figure out the cells' playbook. From there, we can start trying to predict how any individual patient's cancer might respond to one particular treatment or another.

First, we need to measure what's present in each person's cancer, mapping out all the different types of cells, including immune and supporting cells, and also get an idea of the molecular habitat. Then we need to work out how they'll respond to different treatments and pick the one most likely to have the greatest impact. We need to watch what happens and learn from it, figuring out whether or not cancers with similar genetic profiles tend to bounce back in the same way and seeing how predictable this is. Finally, we should be able to work out the best second-line treatment, and the one after that and the one after that. Or maybe we can discover the pharmacological equivalent of Gatenby's office bin, blocking off the routes to resistance. Once we can do that, then we'll have put ourselves firmly in the evolutionary driving seat, steering the disease toward our desired outcome.

Cancer is a complex evolving system, making the game we're playing more like chess than a simple office chase, but we should still be able to work out how to win. There's a huge number of possible configurations of pieces once the game is going, but the rules are always the same. Bishops move diagonally. Knights do that jumpy thing. Queens can go in any direction. Each game is unique—just as every cancer is different—from the choice of opening gambit to the final configuration of checkmate, so mindlessly pulling out the same sequence of moves is never going to work.

Luckily, we have a major strategic advantage: intelligence. With the exception of contagious cancers like the Tasmanian devil facial tumors, each individual person's cancer can't pass on the evolutionary tricks it has learned to the next generation. It dies along with the body of its host. But we can learn from every single case to figure out exactly how and why things went wrong, and try something different next turn.

We'll probably have to adjust our strategy in response to any unexpected moves once the game is in play, but if we know the rulebook inside out, we can become grandmasters, thinking five or six moves ahead. And eventually, because each individual cancer is a one-off event—a newbie coming in cold to the game—we should be able to win every time.

## LEARNING FROM FAILURE

As far as Bob Gatenby is concerned, a greater understanding of these evolutionary games urgently needs to become an integral part of the drug development and regulatory landscape, instead of bringing ever more "magic bullets" to market without a thought for what happens when they inevitably fail. When pesticide companies want to bring a new product to market, they're also obliged to file a resistance management program, identifying possible mechanisms of resistance and how to avoid them in order to get it approved. So, Gatenby argues, why isn't this part of the testing and approval process for cancer drugs, too?

The problem is, understanding and addressing drug resistance and tumor evolution in this way means that cancer researchers and doctors need to face the fact that treatments for advanced cancer don't work and figure out why. While nobody likes to look back at their failures, least of all doctors and pharmaceutical company executives, refusing to engage with this concept makes no sense when compared with other high-stakes industries with life or death outcomes. Just think about the amount of investigation that goes on every time a plane crashes: fleets are grounded, black boxes are rescued and pored over to understand what went wrong, and safety features are put in place to ensure it doesn't happen again.

There are several things that help to explain this unwillingness to step back and truly analyze why something went wrong when someone dies from cancer. Unlike the aftermath of air crashes, where grieving families and shocked governments demand action, it's maybe seen as unseemly to clamor for an investigation into what happens when cancer treatment fails. Despite the fact that therapies for advanced cancer cost so much for relatively modest benefits, the incentives for the pharmaceutical industry and regulatory agencies are all stacked toward coming up with new solutions, rather than trying to figure out why their current fixes don't seem to be providing the long-term survival benefits that patients so desperately need.

From a doctor's perspective, there may be fears of criticism and blame—Did you make the wrong decisions? Could you have done something differently?—fueled by an increasingly litigious culture that seizes on any hint of medical negligence. This is balanced by paternalism and hubris in the medical profession, insisting that they gave it their best shot and it's nobody's fault that it didn't work. And then there's a sense of fatalism, especially for cancers where long-term survival remains stubbornly low. Everyone wants to "move the needle," as the saying goes, but if a doubling of survival still means that 90 percent of patients will die within a couple of years, it may be hard to believe that anything will make a meaningful difference. None of these attitudes is particularly helpful for making progress in understanding how to treat people better in the future.

The final hurdle is the most painful to contemplate. In order to figure out why and how someone's cancer evolved resistance to therapy and ultimately killed them, scientists need samples from those tumors at the end of the journey. Death of a loved one from cancer brings many feelings for those who are left behind—sadness, loss, anger, frustration, even relief that there's no more suffering. There are practical matters to be dealt with alongside the emotional workload of grief, so it's not surprising that the desire to participate in a research project probably comes a very long way down the list. It's rare that tumors will be collected through an autopsy if it's obvious that someone has died of cancer, but this material is absolutely vital if we're to understand the evolutionary playbook from start to finish.

Things are slowly starting to change. Researchers are becoming more willing to start the difficult conversations with patients and their families, and funding organizations are now prepared to stump up the cash for the kind of studies that might have previously been seen as too morbid and depressing to contemplate. At University College London, Dr. Mariam Jamal-Hanjani is running a unique trial called Posthumous Evaluation of Advanced Cancer Environment (PEACE)—a tortuous name devised to create the perfect acronym—gathering tumor samples from patients once they've died in order to figure out the final stages of the disease's evolutionary journey. PEACE grew out of the TRACERx lung cancer evolution study (page 173), which she's also helping to run. Patients taking part in TRACERx were so used to the idea of giving regular tumor samples throughout the course of their disease that more and more of them were asking Jamal-Hanjani how they could continue to help with the research once they'd passed away.

The study is aiming to recruit 500 patients while they're very much still alive—either people with brain tumors or those with any type of cancer that has spread through their body. Jamal-Hanjani and her team of specialist pathologists are gathering blood samples at regular intervals to look for cancer cells and extract DNA from them. They're also gaining consent for tumor and healthy tissue samples to be collected immediately after death. Similar "warm autopsy" programs are also under way at a small but growing number of centers across the USA and other countries. More than 150 people have volunteered to take part in PEACE so far, often with great enthusiasm. The main resistance has been from squeamish clinical colleagues who feel that the study is too sensitive and controversial, with the typical response from patients being, "It's a no-brainer, Mariam—when we're dead you can take as much tissue as you want."

It's time for the field of oncology—and society more widely—to face up to the fact that unless we can find the courage to understand how we've failed, we're never going to gain the knowledge we so desperately need to win the evolutionary game against cancer in the future.

# 11

# GAME OVER

It's 9 a.m. on a chilly morning in April 2018 and I've come to help poison my friend Tamsin. After checking into the University College Hospital Macmillan Cancer Center (UCLH) in London and getting a few tests, we wait until a chemotherapy chair is free. I attempt to distract her for a couple of hours, pretending not to watch the chemicals slithering down the tube running from a fat plastic bag into the port embedded in her chest. The drugs are sensitive to light, so the pouch is covered with a bright orange sleeve. It's exactly the same color as the emergency survival bag I take on hiking holidays. And, in many ways, this is an emergency in slow motion.

Tamsin was diagnosed with bowel cancer just after Christmas 2017, around the time I started working on this book. After surgery and a few rounds of chemo her outlook is good, even though the rogue cells were on the verge of spreading through her body. She's a brilliant climate scientist who specializes in running computer models designed to predict what might happen to the planet in the future, so she knows her own personal probabilities inside out. According to her calculations, her advantages of being young, fit, and healthy mean that her cancer is about as dangerous as taking a regular hike up Mount Everest, with her odds improving with every passing year. That's cold comfort when you live in the city and prefer cocktails to crampons, but this is the trek she's been sent on.

A year on and she's tired but growing stronger, and we're all full of hope that this was just a Bad Patch that will soon be a distant memory. But the more I speak to people like Bob Gatenby and others working on the evolutionary nature of cancer, the same question comes up in my mind again and again: are we doing the right thing? Arguably, the answer is yes—today's oncologists are using the tools that they've got in the best way that they know how, based on the clinical trial data that's available to them.

The collective efforts of scientists and doctors over the past century have got us to a point where half of all people diagnosed with cancer in the UK will survive for at least ten years. But I have come to believe that we need a complete shift in how we think about the origins, prevention, and treatment of cancer, from its earliest evolution as a scrappy expansionist clone to a malignant Darwinian monster, if we're going to fill that half-empty glass to the brim.

Prevention must be our first priority, rather than being relegated to the bottom of the funding heap. Next comes early diagnosis, aiming to spot troublesome clones at a point when surgery and minimal treatment will be curative. This has to go hand in hand with the development of tests that can tell the bad cells from the sad—simply seeking out every lump or bump without figuring out whether or not it's actually dangerous is a futile strategy. Finally, we need more effective long-term treatments for the advanced cancers that manage to slip through the net.

I've sat through talk after talk at scientific conferences, watching researchers present the same story time after time: we tried a big dose of this drug, the cancer came back; we tried a big dose of that one, it came back; we tried a big dose of our new fancy therapy and . . . the cancer came back. Every single downwards tick on their graph represents a human life that has been lost, marking our collective failure to combat the emergence of drug resistance.

It's baffling to me that more people in the field of oncology haven't noticed the parallels with what's going on in the wider world of ecology and evolution. Perhaps there's a certain snobbishness from the white-coated scientists in their pristine laboratories believing that there's little to learn from ecologists who like to get their boots muddy. But their experience tells us that the secret to controlling advanced cancer in the long term lies not just in expecting resistance, but in actively planning for and managing it too.

Humans are a fundamentally optimistic species. We want to believe that we're doing our best and making a difference, and that doing something is better than nothing. But chemotherapy and targeted treatments place an acute selective pressure on highly aggressive, rapidly evolving cancers, potentially making them even worse. Until we can figure out more effective ways to tame the most dangerous selfish monsters that have already spread far and wide through the body, we may have to accept that sometimes the best treatment may be no treatment at all.

There's growing evidence to suggest that opting for palliative treatments that relieve symptoms and ease pain actually leads to longer survival and a better quality of life for people with very little time left, compared with pursuing last-ditch "curative" treatments like high-dose chemotherapy or expensive targeted therapies. This is particularly relevant in the light of a 2013 study from researchers at the Dana-Farber Cancer Institute in Boston, Massachusetts, which showed that the majority of patients with very advanced cancer don't understand that intensive treatment is not likely to cure them. Wherever it comes from—doctors, the pharmaceutical industry, the media, the Internet—the seductive hope of a miracle cure is misleading people at the most vulnerable point in their lives.

It's not being pessimistic—it's being realistic.

The paradigm of precision oncology and maximum tolerated doses has failed to deliver the improvements in long-term survival from advanced cancer that we so urgently need. And the somatic mutation theory of cancer starts to fall apart once it becomes clear that even healthy tissues are a mess of mutations by middle age. It's time for a new way of thinking about cancer that is embedded in an evolutionary mindset, viewing the disease as a complex, evolving system that emerges from within the shifting environment of the body. Rather than focusing on shopping lists of mutations and molecular targets, we should draw inspiration from the work being done to model other complex natural systems like climate. And we have to accept that we can no more treat cancer with a magic bullet than use a shotgun to stop a hurricane.

## A CHANGE IN THE WEATHER

In the summer of 2016, a group of researchers sat down in a fancy conference center deep in the flat expanse of the Cambridgeshire countryside in the UK. Within sight of the Sanger Institute and its enormous banks of tireless DNA sequencing machines, busily reading the genomes of thousands of tumor samples collected from patients all over the world, the scientists began to draw up the principles of a new way to think and talk about cancer. This eventually coalesced into a hefty treatise published toward the end of 2017, outlining how the principles of evolution and ecology should be applied to treat cancer in a better way.

The idea is to classify tumors not just according to the presence of particular mutations or the place in the body where they started, but also by what's called the Eco-Evo index—a measure of how quickly the cancer cells are evolving and the lushness of the internal habitat in which they're growing. The Evo part takes into account the extent of heterogeneity within a tumor—whether its genetic patchwork is just a couple of large swathes or a scattered mess of unique clones—and how quickly it's changing over time. Is it a system of slowly expanding states, or a rapidly shifting gang warfare situation? Is there a slow and steady tick of Darwinian evolution, fueled by a sedate rate of mutations, or a chaotic explosion of shattered chromosomes and selfish monsters?

These questions can be answered with detailed DNA sequencing and other tests that take the three-dimensional organization of cells within a tumor into account, with regular retesting to see how things are changing. One day it might even be possible to predict the evolutionary capacity of a tumor from a single initial sample, based on the kinds of mutations and patterns that are present within it. Ecology is harder to measure, but it's no less an essential part of the equation. Are tumor cells growing in a barren landscape, low in nutrients and roamed by fierce immune predators, with stiff competition from healthy cells? Or are they bedding down in a rich, toxic swamp in which only the weirdest will survive?

Altogether, the Eco-Evo classification divides cancers into sixteen types, based on all the possible combinations of high or low heterogeneity, fast or slow mutation rate, rich or poor resource availability, and hazardous or safe in terms of immune predators and other threats. Some combinations might be unlikely to exist in real life, but considering the outcomes of each is a useful tool for understanding how any given cancer might behave based on these measurements.

The tumors that score lowest on all counts are like deserts, with few resources and little diversity. Consequently, life can't flourish and evolve. Right over the other end of the scale are cancers that are like resource-rich rainforests, populated with diverse and rapidly changing cellular species. New clones are constantly emerging and being snuffed out in the face of heavy predation by immune cells. Somewhere in the middle, with rich diversity and resources but low predation and evolutionary capacity, are cancers that are like a neatly tended garden, supporting a variety of well-fed species that are protected from predators and change little over time.

Most usefully, the Eco-Evo index also points toward the best approach for treating each of the sixteen types. Cancers with very low diversity, low evolvability, and targetable driver mutations in all of their cells might only need one or two targeted therapies to knock them out completely. Others may be driven to extinction with the careful application of the right selective pressures in the right order, or make easy prey for immunotherapy. Rapidly evolving tumors with a lot of genetic diversity would probably be the best candidates for adaptive therapy, aiming for long-term control rather than cure right from the start. Others might benefit from "ecotherapy," draining the toxic swamp and improving the micro-environment to make it less amenable to cancer cells.

Getting this right isn't going to be easy and if we want to make accurate predictions about the behavior of an individual cancer, we're going to need a *lot* of data. It also needs to be the *right* data, capturing the dimensions of space and time, gathered in a way that preserves information about three-dimensional organization within tumors, and collected at regular time points. Not just easy genetic data, but holistic information about phenotype, immune cells, the state of the micro-environment and the rest of the body too, which can be crunched through the kind of sophisticated algorithms and models that are used to understand and predict the outcomes of other complex systems, such as climate.

For an embarrassingly long time I thought mathematical modelling was basically just building fancy animations of biological processes. In fact, it's about taking a bunch of inputs based on measurements made on real-life cells and cancers—such as the number of cells in a tumor, how fast they multiply, how often they die and the levels of nutrients around them—then putting them into an equation that says how you think they're going to behave. This formula spits out predictions about what will happen over a given period of time, based on whatever starting conditions and other parameters you choose to set. Then you ask: do the predictions of the model match up to what we observe in reality?

If not, then you need to make your model more sophisticated until it starts spitting out realistic results. Perhaps you need to take into account the fact that cells might be moving around as well as proliferating, or that the levels of oxygen in the middle of a tumor are much lower than on the edges. If the predictions are correct, then great—you've effectively built a computer simulation of real life and can start doing "experiments" with it. You can tweak the number of cells you start with, fiddle with the nutrient levels or increase the death rate to mimic the effects of chucking in a cell-killing drug, and then see what happens. If the growth of your virtual tumor slows down or stops completely, that suggests you've got a potentially useful approach to test in a clinical trial.

Personalizing this kind of model by inputting data from an individual patient could be a powerful way of predicting the best way to treat their cancer in an evolution-appropriate way, or at least coming up with a more rational place to start. And rather than running endless simulations to hit on the best way of tackling a particular tumor, there's also an opportunity to harness the talents of people who spend their time plotting the downfall of their enemies for fun: gamers. Similar to other crowd-sourced science projects that have turned lab data into computer games to search for DNA changes in cancer cells or hunt for new galaxies, getting smart people to come up with strategic combinations of weapons and spells to combat a shadowy enemy could be a good source of bright ideas. I love the idea of a teenage gamer sitting in their bedroom in Manchester coming up with a treatment schedule for a pensioner in Florida.

Nice metaphors, computer games, and consensus statements are all very well, but this evolutionary stuff really needs to work and it still needs to prove its value by improving long-term survival. Bob Gatenby's prostate cancer trial is a promising start (page 310), but one trial in one tumor type isn't enough. More are in the pipeline, but they'll take time and money. And just as there are climate skeptics, who dismiss the outputs of the complex models and simulations that researchers like my friend Tamsin dedicate their careers to working out, there are some people who doubt that these Eco-Evo ideas will make much difference to how people are treated. Then there are the voices of the pharmaceutical industry, who would rather get another billion-dollar kinase inhibitor approved than come up with a plan to combat resistance to the ones they've already got.

Partly, this is the influence of five decades of gene-centric research that has tended to view the body as a machine and the genes and molecules within our cells as components analogous to an electronic circuit or computer code. This connection was made as the molecular biology revolution of the 1960s coincided with the rise of consumer electronics and computing, and has been remarkably hard to shake off. Ecologists and evolutionary biologists were dismissed as being irrelevant to the inner workings of the body, while cell biologists and physiologists lost out in favor of the somatic mutation theory and the subsequent genetic gold rush.

As Israeli biochemist Isaac Berenblum wrote back in 1974:

> We find ourselves at the present time in the era of molecular biology, and we are perhaps unduly influenced by the genetic code as the dominant principle in biology. Perhaps, in a decade or two from now, the dominant principle may shift to another plane, which in turn will influence our speculations about tumor causation.

More than forty years on it looks like things are finally starting to change.

I can't write a book like this without referring to geneticist Theodosius Dobzhansky's famous quote, "Nothing in biology makes sense except in the light of evolution." First uttered during a lecture at the 1972 convention of the

National Association of Biology Teachers in the USA, Dobzhansky began his talk by pointing out that in 1966, Sheik Abd al Aziz bin Bad wrote to the King of Saudi Arabia asking him to suppress the terrible heresy that was sweeping the land claiming that the Earth went around the sun and not the other way round. Copernicus got that one nailed in the sixteenth century, so Dobzhansky doubted that the good Sheik could simply be ignorant of the evidence from multiple astronomers and physicists showing that, indeed, *eppur si muove* (". . . and yet it moves"). Dobzhansky concluded, "Even more likely, he is so hopelessly biased that no amount of evidence would impress him. Anyway, it would be a sheer waste of time to attempt to convince him."

Similarly, nothing in cancer makes sense except in the light of evolution. Failing to acknowledge this simple but inconvenient fact of life is the reason why survival from advanced metastatic cancer has barely changed over the years. And unless we really start to get to grips with the underlying evolutionary nature of the disease, I don't believe we're going to get much further. It's a process that has always been at work throughout the history of life on Earth: ignoring the growing amount of evidence showing that it also functions at the level of the cells in our bodies is fast becoming tantamount to heresy.

There are many unanswered questions and a lot of work still to be done, but as Dobzhansky notes:

> Evolution as a process that has always gone on in the history of the earth can be doubted only by those

> who are ignorant of the evidence or are resistant to evidence, owing to emotional blocks or to plain bigotry . . . Imagine that everything is completely known and that science has nothing more to discover: what a nightmare!

## CHASING THE CURE

There's an apocryphal story told by oncologists about a doctor who has just finished treating two women for breast cancer. One of them comes to see him and asks, "Am I cured?" He tells her, "Well, we cannot say you are cured. It may come back in five, ten, twenty years." She starts crying and says, "What do you mean? I have had all the treatment. I have lost my breast. I have lost my job. I have lost my friends. And now you're telling me that I'm not even cured?"

Later that day the other woman comes in and asks the same question: "Am I cured?" Mindful of what had happened earlier, he says, "Yes, you can consider yourself cured." Again, she starts crying and says, "What do you mean? I have lost my breast. I have been through chemotherapy. I have lost my job. I have lost my husband. I will never be the same again. How can you say that I am cured?"

The final shift that has to happen in the move toward a more evolutionary approach to treating cancer is a psychological one. Anyone who has survived cancer knows that it never leaves you. You are forever changed. Years after the shock of diagnosis has receded, the fear of relapse is always there. Some people wear the fear like an animal draped around their shoulders, learning to coexist and carry its weight. Others carry it deep inside, occasionally letting it out late at night for a runaround.

If adaptive therapy or similar long-term control strategies start to become commonplace for advanced cancers, then our relationship to tumors growing inside our bodies has to change, too. The patients on Bob Gatenby's trial had to learn to live with the knowledge that their disease was always there and was deliberately being allowed to grow back. There will need to be a shift in language we use, thinking of these treatments as more like tending a garden and keeping everything as tidy as possible—weeding flower beds and clipping hedges—rather than burning it all down and praying that nothing grows back through the scorched earth.

This psychological change comes into cancer prevention too, through a more nuanced understanding of the interplay between mutations and micro-environment. We either tend to want to pin the blame on specific causes (especially if they're avoidable) and preach about the importance of a "healthy lifestyle," or simply shrug and put it down to fate. Cancer cells and the mutations within them are only half of the picture. Deciphering the mutational fingerprints within tumors presents genuine options for cancer prevention by minimizing exposure to DNA-damaging agents (and potential censure for organizations that deliberately fail to control known carcinogens). But it's impossible to halt the damage that our own cells bring upon themselves, thanks to the fundamental biochemistry of life.

Healthy cells and their habitats are an equally important

component, and maybe we could focus more on the bigger picture of maintaining tissue health and controlling long-term chronic inflammation to keep order in the society of our cells rather than reeling off a list of things to avoid. We should be celebrating the incredible ability of our bodies to suppress cheating cells for as long as they do and not castigate ourselves when one slips through the net. As early diagnosis and treatment improves, perhaps we can even come to see cancer less as the angry hand of fate and more as a normal part of aging—a stage of life that many people will go through. First period, first gray hair, first wrinkles, first clonal expansion.

Declaring war on a biological concept isn't exactly the dumbest idea a politician has ever had, but Nixon's 1971 commencement of the War on Cancer lulled us into the belief that it was something we could win. Instead, it's become an intractable mess, with casualties piling up and ever more money vanishing into the oncogene-pharmaceutical industrial complex. Three decades later, Andrew von Eschenbach, then director of the US National Cancer Institute (NCI), announced in 2003 that the goal of the Institute would be to "eliminate suffering and death due to cancer" by 2015. Yet, despite the NCI's efforts (along with those of thousands of others around the world), people are still suffering and dying as we head into the 2020s.

As someone who spent more than a decade of their life working for a cancer research charity, I appreciate the need for aspirational slogans and motivational big ideas. But repeatedly overpromising and under-delivering is a recipe for public disappointment and disillusionment, creating an environment in which conspiracies and quackery thrive. And putting a hard deadline on this kind of thing is pure magical thinking. However, when anyone does dare to point out that things might be more complicated than we'd initially thought, there is reluctance to accept the truth.

When Mel Greaves established the Center for Evolution and Cancer at the Institute of Cancer Research in London in 2014, he held a press briefing to explain why the realities of natural selection mean that we might never be able to cure advanced cancer owing to the inevitable emergence of resistance. He painted an alternative vision of turning currently fatal metastatic tumors into long-term chronic conditions, where survival is measured in years rather than months, by understanding and steering the evolutionary forces at work.

The next day, a sniffy editorial in *The Times* lambasted Greaves for his lack of aspiration and inspiration, saying:

> Take Cancer Research UK. The charity's slogan is "Together We Will Beat Cancer." If its slogan were, as Professor Greaves might have it, "Together We Will Delay Cancer," would it . . . have raised enough money to fund research into more than 200 different types of cancer? Almost certainly not.

But what should this new slogan be? What should be our aspiration for tackling cancer in the light of the realities of multicellularity and evolution? Throughout the process of writing this book I've spoken with more than fifty thoughtful researchers and ploughed through countless books and papers. I've come to realize that the best way of phrasing this goal is in the words of Peter Campbell, one of the world's leading cancer geneticists at the Sanger Institute, who told me, "What's the endgame? I guess it's that you live long enough to die of something else first."

Real life isn't a mythical epic or fairy tale. Everyone dies in the end. Only gods are immortal—and they don't exist. Our aim should be for everyone to spend their time on this planet in good health for as long as they wish; that has to be what it's all about: preventing anyone and everyone from succumbing to cancer before their time, whatever their age. And if more and more people are going to be living for decades after diagnosis, we need to do much more to mitigate the side effects of treatment and support their physical and mental wellbeing.

Although humanity has a 100 percent mortality rate, life itself keeps going. Cells will continue to proliferate, keeping the unbroken biological link stretching all the way back to LUCA. Cancer is the price of life. We can no more declare war on it than we can declare war on multicellularity or evolution. Without the dangerous genes that drive one cell to become many, we would never grow from one cell into a baby in the uterus or be able to repair and replace our aging parts. Without multicellularity, we'd still be single-celled blobs absentmindedly swimming around in the primordial soup. But the rules that make multicellularity work form the well-behaved society from which rebellious, cheating cells will inevitably emerge.

Without evolution, we and the rest of the gloriously diverse organisms on this planet wouldn't exist at all. Cancer takes the most creative force in nature and uses it to wreak the most terrible destruction. Yet it cannot plan for the future—every tumor is a new evolutionary experiment from which we can learn, turning that knowledge to our advantage.

We need to tell a new story about cancer—not as something alien that we need to nuke from orbit, but as an intrinsic property of multicellular life. We have to understand it from the perspective of evolution and the ecology of the landscape within the body. We have to eradicate it if possible: steering cancer into an evolutionary dead end where every single cell has exhausted all the possible evolutionary avenues and extinction is the only option. Where that's not possible, the alternative is to keep chasing tumors round in circles—watching, waiting, treating, watching, waiting, treating . . . a situation that could continue for decades. And although that may not be the Cure for Cancer that we thought we were searching for, it's something that looks very much like it.

# ACKNOWLEDGMENTS

First and foremost I'd like to thank my agent, Chris Wellbelove at Aitken Alexander, who has supported me every step of the journey from initial proposal to final copy, and to Alison Lewis at The Zoë Pagnamenta Agency for representing me in the US. I owe a big debt of thanks to everyone who's had their editorial fingers on the text, making it a much better (albeit less sweary) book in the process: Jenny Lord, Frank Swain, Maddy Price, and Claire Dean at Weidenfeld & Nicolson in the UK and the team at BenBella Books in the US.

I'm eternally grateful to all my former colleagues at Cancer Research UK—whether in the press and science comms team or the wider organization—for providing inspiration, opportunities, and friendship for more than a decade. Special thanks go to Henry Scowcroft and Ed Yong, my co-conspirators in the early days of the charity's Science Update blog, who did so much to help me hone my writing skills and knowledge.

I am indebted to all the researchers who gave up their time to chat about their work and point me in the right direction. There isn't room to include all their words and stories in this book, but they all helped to shape my thinking in some way:

Alex Cagan, Amy Boddy, Andrea Sottoriva, Andy Futreal, Anna Barker, Anna Trigos, Athena Aktipis, Beata Ujvari, Bissan Al Lazikani, Bob Gatenby, Bob Weinberg, Carlo Maley, Casey Kirkpatrick, Charlie Swanton, Cristian Tomasetti, Daniel Durocher, David Adams, David Basanta, David Goode, Elizabeth Murchison, Fran Balkwill, Frederic Thomas, Gerard Evan, Greg Hannon, Hans Clevers, Hayley Francies, Inaki Ruiz Trillo, Inigo Martincorena, Joel Brown, Kenneth Pienta, Kim Bussey, Kristin Swanson, Manuel Rodrigues, Marc Tollis, Mariam Jamal-Hanjani, Mel Greaves, Mike Stratton, Nicky McGranahan, Olivia Rossanese, Paul and Pauline Davies, Peter Campbell, Phil Jones, Richard Houlston, Richard Peto, Rodrigo Hamede, Ron de Pinho, Rong Li, Ruben van Boxtel, Sam Behjati, Sandy Anderson, Serena Nik-Zainal, Steve Elledge, Steve Jackson, Trevor Graham, Vicky Forster, Walter Bodmer, and Yin-Yin Yuan.

Many thanks to the support staff of all the labs and institutes I visited for their efficiency and kindness in arranging meetings and bringing coffee, particularly the comms teams at the Institute of Cancer Research in Sutton and the Wellcome Sanger Institute in Cambridge. Thanks to Toni Garcia for her assistance in setting up my research trip to the United States and Canada, and to Cyril and Angela Arney and Lucy and Dan Durocher for their kind hospitality while I was passing through.

I'm privileged to be able to share the personal experiences of Josh Barnfather, Crispian Jago, Tamsin Edwards, and Desiree. Thank you for trusting me with your stories.

Huge thanks to the team behind my science communication company, First Create The Media, who kept the wheels turning while I went down the Book Hole, especially our Chief Operations Officer and organizational wizard Sarah Hazell.

Thanks to my family and friends for all their cheerleading, whether in real life or online: Mum, Dad, Lucy, Dan, Chloë, Helen, Rob, and Mattie; Adventure Club (Martin, Jen, Liz, James, and Chris); Smut Club (Safia, Sarah, Aine, Emma, and Nell); The Blue; and my friends and followers on Twitter.

Finally, this book wouldn't have happened without the unwavering love and support of my partner, Martin Robbins, who provided just the right balance of sympathetic Scotch-drinking and kindly arse-kicking. Thank you xx

# GLOSSARY

**Apoptosis:** Controlled cell death (sometimes referred to as "cell suicide" or programmed cell death), which is used to get rid of damaged, old or unwanted cells. Apoptosis is a powerful form of protection against cancer, and tumors often evolve ways of over-riding it.

**Bases/base pairs:** The chemical building blocks that DNA and RNA are made of. There are four bases ("letters"), known as A (adenine), C (cytosine), G (guanine) and T (thymine). A always pairs with T, C always pairs with G, creating the ladder-like structure of DNA.

**Chromosome:** A single long string of DNA.

**Chromothripsis:** Literally "chromosome shattering." Large-scale rearrangement of the DNA within the nucleus of a cell, in which many stretches of DNA are broken up and stuck together in random new configurations.

**Clone:** A group of cells originating from a single founder.

**DNA (deoxyribonucleic acid):** A long ladder-shaped molecule in the shape of a twisted ladder (double helix). The outside struts are a chain of sugary molecules, while the rungs are made from pairs of chemicals called bases. The specific order of these bases carries genetic instructions that the cell uses to make all the molecules of life.

**Driver mutation:** An alteration in an oncogene that enhances the proliferation of cancer cells or confers another kind of competitive advantage.

**Epigenetic:** Factors affecting gene activity that are not directly encoded within DNA itself.

**Extracellular matrix:** The molecular "glue" that helps to hold cells together within the tissues of the body.

**Gene:** A stretch of DNA that carries the information for the cell to make a specific protein or RNA.

**Genome:** The complete set of genetic instructions (DNA) required to build an organism.

**Genotype:** The genetic makeup of an individual cell, tumor or organism.

**Germ cells:** The special cells in an embryo that will eventually make eggs or sperm.

**Histones:** Ball-shaped proteins that package DNA in the nucleus.

**Kinase:** A protein that attaches a chemical "tag," known as a phosphate group, onto another protein. Many kinases are involved in sending signals within and between cells, either telling them to start or stop proliferating.

**Mitosis:** The process of division by which one cell splits into two. Under normal circumstances, each new cell has the same amount of DNA and the same number of chromosomes as the original one. Cancer cells often have faulty mitosis, which can lead to chromosomes being lost or gained in the new progeny.

**Mutation:** An alteration or change in a sequence of DNA. Mutations can occur within genes or in non-coding DNA and can be anything from a single base ("letter") change to large-scale structural rearrangements.

**Natural selection:** First proposed by Charles Darwin, natural selection is the process in which organisms or cells with traits that make them better adapted to their environment are more likely to survive and pass on beneficial genes to their offspring.

**Negative selection:** The evolutionary process by which harmful traits are lost from a population, also known as purifying selection.

**Neutral selection:** The idea that most genetic changes within cells, organisms or populations are neither beneficial or harmful, and therefore are not influenced by positive or negative selection.

**Non-coding DNA:** A stretch of DNA that doesn't carry instructions for making a protein. It might do nothing, or it might be used as a template to make non-coding RNA.

**Nucleus:** The structure within a cell that houses all the DNA. It can be thought of as the "control center" of the cell.

**Oncogene:** A gene encoding a protein that drives cells to multiply. Under normal circumstances, oncogenes maintain the proliferation of new cells only when needed. Over-active oncogenes drive excessive cell growth, potentially leading to cancer.

**Phenotype:** How a cell, tumor or organism looks and behaves.

**Positive selection:** The evolutionary process by which beneficial traits spread through a population.

**Protein:** A molecule made up of a long string of small building blocks called amino acids. Proteins do much of the work in cells, from creating and maintaining structure to carrying out the chemical reactions that keep us alive.

**RNA (ribonucleic acid):** A molecule that is similar to one half of the "ladder" of DNA, produced when a gene is switched on.

**Sequencing:** Reading the order of letters (bases) in any stretch of DNA.

**Soma/Somatic cells:** All the cells or parts of the body except the germ cells.

**Stroma:** The supportive collection of connective tissue, blood vessels, immune cells and extracellular matrix within an organ that is not directly connected to its main function.

**Telomeres:** The molecular "caps" that protect the end of chromosomes.

**Tumor suppressor:** A gene encoding a protein that suppresses the development of cancer, for example by slowing cell proliferation, detecting or repairing genetic damage, or by causing faulty cells to die. Loss of one or more tumor suppressor gene functions is a key step in cancer progression.

# FURTHER READING

There's a lot of useful background about genes and genomes in my first book, *Herding Hemingway's Cats: Understanding How Our Genes Work* (Bloomsbury Sigma, London, 2016).

First published a decade ago, Siddartha Mukherjee's award-winning *The Emperor of all Maladies: A Biography of Cancer* (Scribner, New York, 2010) is a handy overview of the history of cancer research and treatment, although stops short of the most recent advances in genomics. An updated version is due in 2020.

George Johnson weaves together the story of his wife's experience of cancer with fascinating scientific tales in his book *The Cancer Chronicles: Unlocking Medicine's Deepest Mystery* (Penguin Random House, New York, 2013).

In 1951, a young African American woman died of cervical cancer. Today, her cells are grown in labs all over the world. Rebecca Skloot addresses the myths and misconceptions surrounding one of the most important figures in the history of cancer research in *The Immortal Life of Henrietta Lacks* (Crown Publishing Group, New York, 2010).

Jessica Wapner's *The Philadelphia Chromosome: A Mutant Gene and the Quest to Cure Cancer at the Genetic Level* (The Experiment, New York, 2013) explores the story behind the discovery of Glivec—arguably the most successful cancer drug ever invented, which set the paradigm for targeted therapy.

Although it's not strictly about cancer, Jonathan Losos' *Improbable Destinies: Fate, Chance, and the Future of Evolution* (Penguin Random House, New York, 2017) takes a broad look at the ways in which convergent evolution has shaped life on Earth.

Probably the first book to take a look at cancer from an evolutionary perspective, *Cancer: The Evolutionary Legacy* by Melvyn Greaves (Oxford University Press, Oxford, 2000) is a little dated but still contains plenty of insights, while James DeGregori's *Adaptive Oncogenesis* is a more recent overview of current evolutionary thinking around cancer (Harvard University Press, 2018).

*Frontiers in Cancer Research: Evolutionary Foundations, Revolutionary Directions* (Springer-Verlag, New York, 2016) is a compilation of scientific essays edited by Carlo Maley and Mel Greaves, capturing much of the new thinking around cancer development and treatment.

*Ecology and Evolution of Cancer* (Academic Press, Cambridge, Mass., 2017), edited by Beata Ujvari, Benjamin Roche, and Frédéric Thomas, takes a deep dive into the world of tumor evolution, and also contains an impressive list of all the species known to be affected by cancer to date. Frédéric Thomas has also written a French language overview of cancer evolution, L'abominable secret du cancer (HumenSciences, Paris, 2019) aimed at the general public.

Although they were controversial at the time, the ideas put forward by Carlos Sonnenschein and Ana Soto in *The Society of Cells:*

*Cancer Control of Cell Proliferation* (Taylor & Francis, Abingdon, 1999) are becoming more relevant as scientists search for more integrated, tissue-based models of tumor development and growth.

If you're a podcast fan, I recommend Vinay Prasad's Plenary Session podcasts, in which he lets rip at bad health policy, over-hyped cancer treatment, and poorly designed clinical trials on a regular basis. Find them via Twitter (@Plenary_Session) or wherever you get your podcasts.

I also present Genetics Unzipped, a fortnightly podcast from The Genetics Society, which covers modern and historical stories from the world of genes, genomes, and DNA, including cancer. Find it online at GeneticsUnzipped.com, or search your favorite podcast app.

# REFERENCES

## Introduction

*Third Annual Report of the Imperial Cancer Research Fund* (1905), p8
Bailar, J.C. and Smith, E.M. (1986) Progress against cancer? *New England Journal of Medicine* 314:1226–32 doi:10.1056/NEJM198605083141905
Dietrich, M. (2003) Richard Goldschmidt: hopeful monsters and other "heresies," *Nat Rev Genet* 4: 68–74 doi:10.1038/nrg979
Forster, V. (2019) An Israeli Company Claims That They Will Have A Cure For Cancer In A Year. Don't Believe Them, *Forbes* (published online January 30, 2019) bit.ly/2ufqPJs
Power, D'A. (1904) Notes on an ineffectual treatment of cancer: being a record of three cases injected with Dr. Otto Schmidt's serum, *Br Med J.* 1: 299–302 doi:10.1136/bmj.1.2249.299

## Chapter 1

Weiss, M., Sousa, F., Mrnjavac, N. et al. (2016) The physiology and habitat of the last universal common ancestor. Nat Microbiol 1: 16116 doi:10.1038/nmicrobiol.2016.116
Galen, *On the Method of Healing to Glaucon*, 2.12, 11.140–41K
David, A. and Zimmerman, M. (2010) Cancer: an old disease, a new disease or something in between? *Nat Rev Cancer* 10: 728–733 doi:10.1038/nrc2914
Scientists suggest that cancer is man-made (2019) Manchester University website (published online October 14 2019) bit.ly/2sziYpK
Hunt, K., Kirkptarick, C., Campbell, R. and Willoughby, J. Cancer Research in Ancient Bodies (CRAB) Database cancerantiquity.org/crabdatabase
Banks Whitely, C. and Boyer, J.L. (2018) Assessing cancer risk factors faced by an Ancestral Puebloan population in the North American Southwest,

*International Journal of Paleopathology* 21: 166–177 doi:10.1016/j.ijpp.2017.06.004
Buikstra, J.E. and Ubelaker, D.H. (1994) Standards for data collection from human skeletal remains. *Arkansas Archeological Survey Research Series* No. 44 doi:10.1002/ajhb.1310070519
Lynnerup, N. and Rühli, F. (2015) Short review: the use of conventional X rays in mummy studies, *The Anatomical Record* 298: 1085–1087 doi:10.1002/ar.23147
Strouhal E. (1976) Tumors in the remains of ancient Egyptians, *Am J Phys Anthropol.* 45: 613–20 doi:10.1002/ajpa.1330450328
Odes, E.J., Randolph-Quinney, P.S., Steyn, M., et al. (2016) Earliest hominin cancer: 1.7-million-year-old osteosarcoma from Swartkrans Cave, South Africa. *South African Journal of Science* 112: Art. #2015-0471 doi:10.17159/sajs.2016/20150471
Odes, E.J., Delezene, L.K., Randolph-Quinney, P.S. et al. (2018) A case of benign osteogenic tumor in Homo naledi: Evidence for peripheral osteoma in the U.W. 101–1142 mandible, *International Journal of Paleopathology* 21: 47–55 doi:10.1016/j.ijpp.2017.05.003
Czarnetzki, A., Schwaderer, E. and Pusch, C.M. (2003) Fossil record of meningioma, *The Lancet* 362: 408 doi:10.1016/S0140-6736(03)14044-5
Molto, E., Sheldrick, P. (2018) Paleo-oncology in the Dakhleh Oasis, Egypt: Case studies and a paleoepidemiological perspective, *International Journal of Paleopathology* 21:96–110 doi:10.1016/j.ijpp.2018.02.003
Domazet-Lošo, T., Klimovich, A., Anokhin, B. et al. (2014) Naturally occurring tumors in the basal metazoan *Hydra*, *Nat Commun* 5: 4222 doi:10.1038/ncomms5222
Haridy, Y., Witzmann, F., Asbach, P., Schoch, R.R., Fröbisch, N., Rothschild, B.M. (2019) Triassic Cancer—Osteosarcoma in a 240-Million-Year-Old Stem-Turtle. *JAMA Oncol.* 5:425–426. doi:10.1001/jamaoncol.2018.6766
Ujvari, B., Roche, B. and Thomas, F. (2017) *Ecology and Evolution of Cancer*, Academic Press, Cambridge, Mass. Chapter 2.
Shufeldt, R.W. (1919) A three-legged robin (*Planesticus m. migratorius*), *The Auk* 36: 585–586 doi:10.2307/4073388
Rothschild, B.M., Tanke, D.H., Helbling, M. et al. (2003) Epidemiologic study of tumors in dinosaurs. *Naturwissenschaften* 90, 495–500 doi:10.1007/s00114-003-0473-9
Henrique de Souza Barbosa, F., Gomes da Costa Pereira, P.V.L, Paglarelli, L. et al. (2016) Multiple neoplasms in a single sauropod dinosaur from the Upper Cretaceous of Brazil. *Cretaceous Research* 62: 13–17 doi:10.1016/j.cretres.2016.01.010
Brem, H. and Folkman, J. (1975) Inhibition of tumor angiogenesis mediated by cartilage. *J Exp Med* 141: 427–439 doi:10.1084/jem.141.2.427
Main, D. (2013) Sharks Do Get Cancer: Tumor Found in Great White, LiveScience (published online December 3, 2013) bit.ly/2MMrp7V

McInnes, E. F., Ernst, H., and Germann, P.-G. (2013). Spontaneous neoplastic lesions in control Syrian hamsters in 6-, 12-, and 24-month short-term and carcinogenicity studies. *Toxicologic Pathology,* 41(1), 86–97 doi:10.1177/0192623312448938

Henwood, Chris (2001) The Discovery of the Syrian Hamster, *Mesocricetus auratus*, *The Journal of the British Hamster Association* 39 bit.ly/2szzCWh

Gordon, M. (1941) Genetics of melanomas in fishes v. the reappearance of ancestral micromelanophores in offspring of parents lacking these cells, *Cancer Res 1*: 656–659

Munk, B.A., Garrison, E., Clemons, B., & Keel, M.K. (2015). Antleroma in a free-ranging white-tailed deer (*Odocoileus virginianus*). *Veterinary Pathology,* 52: 213–216 doi:10.1177/0300985814528216

Peto R. (2015) Quantitative implications of the approximate irrelevance of mammalian body size and lifespan to lifelong cancer risk. *Phil. Trans. R. Soc.* B 370: 20150198 doi:10.1098/rstb.2015.0198

Fisher, D.O., Dickman, C.R., Jones, M.E., Blomberg, S.P. (2013) Evolution of suicidal reproduction in mammals, *Proc Natl Acad Sci U S A.* 110: 17910–17914 doi:10.1073/pnas.1310691110

Nielsen, J., Hedeholm, R.B., Heinemeier, J. et al. (2016) Eye lens radiocarbon reveals centuries of longevity in the Greenland shark (*Somniosus microcephalus*), *Science* 353:702–4. doi:10.1126/science.aaf1703

Boddy, A.M., Huang, W., Aktipis, A. (2018) Life history trade-offs in tumors, *Curr Pathobiol Rep.* 6: 201–207 doi:10.1007/s40139-018-0188-4

Avivi, A., Ashur-Fabian, O., Joel, A. et al. (2007) P53 in blind subterranean mole rats—loss-of-function versus gain-of-function activities on newly cloned Spalax target genes, *Oncogene* 26: 2507–2512 doi:10.1038/sj.onc.1210045

Domankevich, V., Eddini, H., Odeh, A. and Shams, I. (2018). Resistance to DNA damage and enhanced DNA repair capacity in the hypoxia-tolerant blind mole rat *Spalax carmeli. J. Exp. Biol.* 221: jeb174540 doi:10.1242/jeb.174540

Hilton, H.G., Rubinstein, N.D., Janki, P. et al. (2019) Single-cell transcriptomics of the naked mole-rat reveals unexpected features of mammalian immunity, *PLoS Biol* 17: e3000528 doi:10.1371/journal.pbio.3000528

Seluanov, A., Hine, C., Azpurua, J., et al. (2009) Hypersensitivity to contact inhibition provides a clue to cancer resistance of naked mole-rat, *Proc Natl Acad Sci U S A.* 106:19352-7 doi:10.1073/pnas.0905252106

Herrera-Álvarez, S., Karlsson, E., Ryder, O.A. et al. (2018) How to make a rodent giant: Genomic basis and tradeoffs of gigantism in the capybara, the world's largest rodent, *bioRxiv* 424606; doi:10.1101/424606

Keane, M., Semeiks, J., Webb, A. E. et al. (2015). Insights into the evolution of longevity from the bowhead whale genome. *Cell reports* 10: 112–122 doi:10.1016/j.celrep.2014.12.008

Seim, I., Fang, X., Xiong, Z. et al. (2013) Genome analysis reveals insights into physiology and longevity of the Brandt's bat *Myotis brandtii. Nat Commun* 4: 2212 doi:10.1038/ncomms3212

Nagy, J.D., Victor, E.M., Cropper, J.H. (2007) Why don't all whales have cancer? A novel hypothesis resolving Peto's paradox. *Integr Comp Biol.* 47:317-28. doi:10.1093/icb/icm062

Cancer risk statistics, Cancer Research UK website cancerresearchuk.org/health-professional/cancer-statistics/risk

Crain, Esther, "These Are the Odds You'll Get Cancer During Your Lifetime," https://www.womenshealthmag.com/health/a19902127/chances-of-cancer/

## Chapter 2

Karpinets, T., Greenwood, D. J., Pogribny, I., and Samatova, N. (2006) Bacterial stationary-state mutagenesis and Mammalian tumorigenesis as stress-induced cellular adaptations and the role of epigenetics, *Current Genomics* 7: 481–496 doi:10.2174/138920206779315764

Buss L.W. (1982) Somatic cell parasitism and the evolution of somatic tissue compatibility, *Proceedings of the National Academy of Sciences USA 79*: 5337–5341 doi:10.1073/pnas.79.17.5337

Santorelli, L., Thompson, C., Villegas, E. et al. (2008) Facultative cheater mutants reveal the genetic complexity of cooperation in social amoebae, *Nature* 451: 1107–1110 doi:10.1038/nature06558

Khare, A. and Shaulsky, G. (2010) Cheating by Exploitation of Developmental Prestalk Patterning in *Dictyostelium discoideum*, *PLoS Genet* 6: e1000854 doi:10.1371/journal.pgen.1000854

Strassmann, J.E., Zhu, Y. and Queller, D.C. (2000) Altruism and social cheating in the social amoeba Dictyostelium discoideum, *Nature* 408: 965–7 doi:10.1038/35050087

Santorelli, L.A., Kuspa, A., Shaulsky, G. et al. (2013) A new social gene in *Dictyostelium discoideum*, chtB, *BMC Evol Biol* 13: 4 doi:10.1186/1471-2148-13-4

Cherfas, J. (1977) The Games Animals Play, *New Scientist* 75: 672–673

Collins, J. (2014) The origin of the phrase "sneaky f**cker", Jason Collins blog (published online January 8, 2014) bit.ly/2ZTrQ5B

Aumer, D., Stolle, E., Allsopp, M. et al. (2019) A single SNP turns a social honey bee (*Apis mellifera*) worker into a selfish parasite, *Molecular Biology and Evolution* 36: 516–526 doi:10.1093/molbev/msy232

Aktipis A. (2015). Principles of cooperation across systems: from human sharing to multicellularity and cancer, *Evolutionary Applications* 9: 17–36. doi:10.1111/eva.12303

Sorkin, R.D. (2000) A Historical Perspective on Cancer, *arXiv* (submitted 1 November 2000) arxiv.org/abs/physics/0011002

Davies, P. C., & Lineweaver, C. H. (2011). Cancer tumors as Metazoa 1.0: tapping genes of ancient ancestors, *Physical Biology* 8: 015001 doi:10.1088/1478-3975/8/1/015001

Munroe, R. Physicists, *XKCD* xkcd.com/793/
Trigos, A.S., Pearson, R.B., Papenfuss, A.T. and Goode, D.L. (2017) Atavistic gene expression patterns in solid tumors, *Proceedings of the National Academy of Sciences USA* 114: 6406–6411 doi:10.1073/pnas.1617743114
Trigos, A.S., Pearson, R.B., Papenfuss, A.T. and Goode, D.L. (2019) Somatic mutations in early metazoan genes disrupt regulatory links between unicellular and multicellular genes in cancer, *eLife* 8: e40947 doi:10.7554/eLife.40947

## Chapter 3

Parts of this chapter are adapted from my feature "The DNA detectives hunting the causes of cancer, published by Wellcome on Mosaic, reproduced here under a Creative Commons license (published online September 25, 2018) bit.ly/DNADetectives
Faguet, G.B. (2014) A brief history of cancer: Age␦old milestones underlying our current knowledge database, *Int J Cancer* 136: 2022–2036 doi:10.1002/ijc.29134
Hadju, S.I. (2006) Thoughts about the cause of cancer, *Cancer* 8: 1643–1649 doi:10.1002/cncr.21807
Scowcroft, H. (2008) Is this the start of the silly season? Cancer Research UK Science blog (published online July 11, 2008) bit.ly/39DNOxN
Scowcroft, H. (2011) No need to worry about having a shower or drinking water. Cancer Research UK Science blog (published online March 17, 2011) bit.ly/2sHUASA
Turning on the light to go to the toilet does not give you cancer. University of Leicester website (published online April 14, 2010) bit.ly/35na8bP
Emami, S. A., Sahebkar, A., Tayarani-Najaran, N., and Tayarani-Najaran, Z. (2012) Cancer and its Treatment in Main Ancient Books of Islamic Iranian Traditional Medicine (7th to 14th Century AD), *Iranian Red Crescent Medical Journal* 14: 747–757 doi:10.5812/ircmj.4954
Triolo, V.A. (1965) Nineteenth century foundations of cancer research advances in tumor pathology, nomenclature, and theories of oncogenesis, *Cancer Res.* 25: 75–106
Triolo, V.A. (1964) Nineteenth century foundations of cancer research origins of experimental research, *Cancer Res.* 24: 4–27
Paweletz, N. (2001) Walther Flemming: pioneer of mitosis research, *Nat Rev Mol Cell Biol* 2: 72–75 doi:10.1038/35048077
Wunderlich, V. (2007) Early references to the mutational origin of cancer, *International Journal of Epidemiology* 36: 246–247 doi:10.1093/ije/dyl272
Hill, J. (1761) *Cautions against the immoderate use of snuff. Founded on the known qualities of the tobacco plant and the effects it must produce when this way taken into the body and enforced by instances of persons who have perished miserably of diseases, occasioned, or rendered incurable by its use*, R. Baldwin and J. Jackson bit.ly/2ZP5wKq

Pott, P. (1775) *Chirurgical observations: relative to the cataract, the polypus of the nose, the cancer of the scrotum, the different kinds of ruptures, and the mortification of the toes and feet*, L. Hawes, W. Clarke, and R. Collins bit. ly/2FkrX0K

Butlin, H.T. (1892) Three Lectures on Cancer of the Scrotum in Chimney-Sweeps and Others: Delivered at the Royal College of Surgeons of England, *Br Med J.* 2: 66-71 doi:10.1136/bmj.2.1645.66

Herr, H.W. (2011) Percival Pott, the environment and cancer, *BJU International* 108: 479–481 doi:10.1111/j.1464-410X.2011.10487.x

Passey, R.D. and Carter-Braine, J. (1925) Experimental soot cancer, *The Journal of Pathology and Bacteriology* 28: 133-144 doi:/10.1002/path.1700280202

Kennaway E.L. (1930) Further experiments on cancer-producing substances, *The Biochemical Journal* 24: 497–504 doi:10.1042/bj0240497

Doll, R. and Hill, A.B. (1950) Smoking and carcinoma of the lung; preliminary report, *British Medical Journal* 2: 739–748. doi:10.1136/bmj.2.4682.739

Proctor, R.N. (2006) Angel H. Roffo: the forgotten father of experimental tobacco carcinogenesis, *Bulletin of the World Health Organization* 84: 494–496 doi:10.2471/blt.06.031682

Doll, R. (1999) Tobacco: a medical history, *Journal of Urban Health* 76: 289–313 doi:10.1007/BF02345669 Proctor, R.N. (2001) Commentary: Schairer and Schöniger's forgotten tobacco epidemiology and the Nazi quest for racial purity, *International Journal of Epidemiology* 30: 31–34 doi:10.1093/ije/30.1.31

Pleasance, E.D., Stephens, P.J., O'Meara, S. et al. (2010) A small-cell lung cancer genome with complex signatures of tobacco exposure, *Nature* 463: 184–190 doi:10.1038/nature08629

Pleasance, E.D., Cheetham, R.K., Stephens, P.J. et al. (2010). A comprehensive catalogue of somatic mutations from a human cancer genome, *Nature* 463: 191–196 doi:10.1038/nature08658

Alexandrov, L.B., Ju, Y.S., Haase, K. et al. (2016) Mutational signatures associated with tobacco smoking in human cancer, *Science* 354: 618–622 doi:10.1126/science.aag0299

COSMIC Catalogue of Somatic Mutations in Cancer cancer.sanger.ac.uk/cosmic/signatures

Kucab, J.E., Zou, X., Morganella, S. et al. (2019) A Compendium of Mutational Signatures of Environmental Agents, *Cell* 177: 821–836.E16 doi:10.1016/j.cell.2019.0

Martin, D. (2003) Douglas Herrick, 82, Dies; Father of West's Jackalope, *New York Times* (published January 19, 2003) nyti.ms/2ST9Nej

Rubin, H. (2011) The early history of tumor virology: Rous, RIF, and RAV, *Proceedings of the National Academy of Sciences USA* 108: 14389–14396 doi:10.1073/pnas.1108655108

Javier, R.T. and Butel, J.S. (2008) The History of Tumor Virology, *Cancer Res* 68: 7693–7706 doi:10.1158/0008-5472.CAN-08-3301

## Chapter 4

Duesberg, P.H. and Vogt, P.K. (1970) Differences between the Ribonucleic Acids of Transforming and Nontransforming Avian Tumor Viruses, *Proceedings of the National Academy of Sciences USA* 67: 1673–1680 doi:10.1073/pnas.67.4.1673

Bister, K. (2015) Discovery of oncogenes, *Proceedings of the National Academy of Sciences USA* 112: 15259–15260 doi:10.1073/pnas.1521145112

Shih, C., Shilo, B.Z., Goldfarb, M.P., Dannenberg, A. and Weinberg, R.A. (1979) Passage of phenotypes of chemically transformed cells via transfection of DNA and chromatin, *Proceedings of the National Academy of Sciences USA* 76: 5714–5718 doi:10.1073/pnas.76.11.5714

Prior, I. A., Lewis, P. D. and Mattos, C. (2012) A comprehensive survey of Ras mutations in cancer, *Cancer Research* 72: 2457–2467 doi:10.1158/0008-5472.CAN-11-2612

Shih, C. and Weinberg, R.A. (1982) Isolation of a transforming sequence from a human bladder carcinoma cell line, *Cell* 29: 161–169 doi:10.1016/0092-8674(82)90100-3

Harper, P.S. (2006) The discovery of the human chromosome number in Lund, 1955–1956, *Hum Genet.* 119: 226–32 doi:10.1007/s00439-005-0121-x

Van der Groep, P., van der Wall, E., and van Diest, P. J. (2011). Pathology of hereditary breast cancer, *Cellular Oncology* 34: 71–88. doi:10.1007/s13402-011-0010-3

Krush, A. J. (1979) Contributions of Pierre Paul Broca to cancer genetics, *Transactions of the Nebraska Academy of Sciences and Affiliated Societies* 316 digitalcommons.unl.edu/tnas/316/

Ricker, C. (2017) From family syndromes to genes... The first clinical and genetic characterizations of hereditary syndromes predisposing to cancer: what was the beginning? *Revista Médica Clínica Las Condes* 28: 482–490 doi:10.1016/j.rmclc.2017.06.011

McKay, A. (2019) *Daughter of Family G,* Knopf Canada amimckay.com/memoir/

Pieters T. (2017) Aldred Scott Warthin's Family "G": The American Plot Against Cancer and Heredity (1895–1940). In: Petermann H., Harper P., Doetz S. (eds) *History of Human Genetics,* Springer

Nair, V.G. and Krishnaprasad H.V. (2015) Aldred Scott Warthin: Pathologist and teacher par excellence, *Arch Med Health Sci* 5:123–5 doi:10.4103/amhs.amhs_135_16

Lynch, H.T. and Krush, A.J. (1971) Cancer family "G" revisited: 1895-1970, *Cancer* 27: 1505–1511 doi:10.1002/1097-0142

McNeill, L. (2018) The History of Breeding Mice for Science Begins With a Woman in a Barn, *Smithsonian Magazine* (published online March 20, 2018) bit.ly/2QjBRWD

Slye, M. (1922) Biological evidence for the inheritability of cancer in man: studies in the incidence and inheritability of spontaneous tumors in mice: Eighteenth Report, *The Journal of Cancer Research* 7: 107-147 doi:10.1158/jcr.1922.107

Muhlenkamp, K. (2014) Storm Driven, *UChicago Magazine* bit.ly/2QkhOas

Lockhart-Mummery, P. (1925) Cancer and heredity, *The Lancet* 205: 427–429 doi:10.1016/S0140-6736(00)95996-8

Harris, H., Miller, O.J., Klein, G. et al. (1969) Suppression of malignancy by cell fusion, *Nature* 223: 363–8 doi:10.1038/223363a0

Harris, H. (1966) Review Lecture Hybrid cells from mouse and man: a study in genetic regulation, *Proc. R. Soc. Lond. B* 166: 358-368 doi:10.1098/rspb.1966.0104

Knudson A. G. (1971) Mutation and cancer: statistical study of retinoblastoma, *Proceedings of the National Academy of Sciences USA* 68: 820–823 doi:10.1073/pnas.68.4.820

Friend, S., Bernards, R., Rogelj, S. et al. (1986) A human DNA segment with properties of the gene that predisposes to retinoblastoma and osteosarcoma, *Nature* 323: 643–646 doi:10.1038/323643a0

Solomon, E., Voss, R., Hall, V. et al. (1987) Chromosome 5 allele loss in human colorectal carcinomas, *Nature* 328: 616–619 doi:10.1038/328616a0

Fearon, E.R. and Vogelstein, B. (1990) A genetic model for colorectal tumorigenesis, *Cell* 61: 759–767 doi:10.1016/0092-8674(90)90186-I

Hahn, W., Counter, C., Lundberg, A. et al. (1999) Creation of human tumor cells with defined genetic elements, *Nature* 400: 464–468 doi:10.1038/22780

Land, H., Parada, L. and Weinberg, R. (1983) Tumorigenic conversion of primary embryo fibroblasts requires at least two cooperating oncogenes, *Nature* 304: 596–602 doi:10.1038/304596a0

Bailey, M.H., Tokheim, C., Porta-Pardo, E. et al (2018) Comprehensive characterization of cancer driver genes and mutations, *Cell* 173: 371–385.e18 doi:10.1016/j.cell.2018.02.060

Martincorena, I., Raine, K.M., Gerstung, M., Dawson, K.J., Haase, K. et al. (2017) Universal patterns of selection in cancer and somatic tissues, *Cell* 171: 1029–1041.e21 doi:10.1016/j.cell.2017.09.042

Martincorena, I., Roshan, A., Gerstung, M. et al (2015) Tumor evolution. High burden and pervasive positive selection of somatic mutations in normal human skin, *Science* 348: 880–886 doi:10.1126/science.aaa6806

Moore, M.R., Drinkwater, N.R., Miller, E.C. et al. (1981) Quantitative Analysis of the Time-dependent Development of Glucose-6-phosphatase-deficient Foci in the Livers of Mice Treated Neonatally with Diethylnitrosamine, *Cancer Research* 41: 1585–1593

Genovese, G., Kähler, A.K., Handsaker, R.E. et al (2014) Clonal Hematopoiesis and Blood-Cancer Risk Inferred from Blood DNA Sequence, *N Engl J Med* 371: 2477–2487 doi:10.1056/NEJMoa1409405

Murai, K., Skrupskelyte, G., Piedrafita, G. et al (2018) Epidermal tissue adapts to restrain progenitors carrying clonal p53 mutations, *Cell* 23: 687–699.e8 doi:10.1016/j.stem.2018.08.017

Martincorena, I., Fowler, J. C., Wabik, A. et al (2018) Somatic mutant clones colonize the human esophagus with age, *Science* 362: 911–917 doi:10.1126/science.aau3879

Risques, R.A., Kennedy, S.R. (2018) Aging and the rise of somatic cancer-associated mutations in normal tissues, *PLoS Genet* 14: e1007108 doi:10.1371/journal.pgen.1007108

Anglesio, M.S., Papadopoulos, N. Ayhan, A. et al. (2017) Cancer-Associated Mutations in Endometriosis without Cancer, *N Engl J Med* 376: 1835–1848 doi:10.1056/NEJMoa1614814

García-Nieto, P.E., Morrison, A.J. and Fraser, H.B. (2019) The somatic mutation landscape of the human body, *Genome Biol* 20: 298 doi:10.1186/s13059-019-1919-5

## Chapter 5

Rich, A.R. (2007) On the frequency of occurrence of occult carcinoma of the prostate, *International Journal of Epidemiology* 36: 274–277 doi:10.1093/ije/dym050

Folkman, J., Kalluri, R. (2004) Cancer without disease, *Nature* 427: 787 doi:10.1038/427787a

Martincorena, I., Raine, K.M., Gerstung, M. et al. (2017) Universal patterns of selection in cancer and somatic tissues, *Cell* 171: 1029–1041.e21 doi: 10.1016/j.cell.2017.09.042

Ecker, B.L., Kaur, A., Douglass, S.M. et al. (2019) Age-Related Changes in HAPLN1 Increase Lymphatic Permeability and Affect Routes of Melanoma Metastasis, *Cancer Discov* 9: 82–95 doi:10.1158/2159-8290.CD-18-0168

Kaur, A., Ecker, B.L., Douglass, S.M. et al. (2019) Remodeling of the Collagen Matrix in Aging Skin Promotes Melanoma Metastasis and Affects Immune Cell Motility, *Cancer Discov* 9: 64–81 doi:10.1158/2159-8290.CD-18-0193

Liu, N., Matsumura, H., Kato, T. et al. (2019) Stem cell competition orchestrates skin homeostasis and ageing, *Nature* 568: 344–350 doi:10.1038/s41586-019-1085-7

Pal, S. and Tyler, J.K. (2016) Epigenetics and aging, *Science Advances* 2: e1600584 doi:10.1126/sciadv.1600584

Raj, A., & van Oudenaarden, A. (2008) Nature, nurture, or chance: stochastic gene expression and its consequences, *Cell* 135: 216–226. doi:10.1016/j.cell.2008.09.050

Watson, C.J., Papula, A., Poon, Y.P.G. et al. (2019) The evolutionary dynamics and fitness landscape of clonal hematopoiesis *bioRxiv* 569566 doi:10.1101/569566

The Great Sausage Duel of 1865 (2014). Skulls in the Stars blog (published online November 1, 2014) bit.ly/39CD1nD

Walter, E., & Scott, M. (2017) The life and work of Rudolf Virchow 1821–1902: "Cell theory, thrombosis and the sausage duel," *Journal of the Intensive Care Society* 18: 234–235 doi:10.1177/1751143716663967

Davillas, A., Benzeval, M., and Kumari, M. (2017). Socio-economic inequalities in C-reactive protein and fibrinogen across the adult age span: Findings from Understanding Society, *Scientific reports* 7: 2641 doi:10.1038/s41598-017-02888-6

Arney, K. (2017) How your blood may predict your future health, *Guardian* (published online October 10, 2017) bit.ly/37AcCoL

Furman, D., Campisi, J., Verdin, E. et al. (2019) Chronic inflammation in the etiology of disease across the life span, *Nat Med* 25: 1822–1832 doi:10.1038/s41591-019-0675-0

Pelosi, A. J. (2019). Personality and fatal diseases: Revisiting a scientific scandal, *Journal of Health Psychology* 24: 421–439 doi:10.1177/1359105318822045

Ana Paula Zen Petisco Fiore, A.P.Z., de Freitas Ribeiro P. and Bruni-Cardoso, A. (2018) Sleeping Beauty and the Microenvironment Enchantment: Microenvironmental Regulation of the Proliferation-Quiescence Decision in Normal Tissues and in Cancer Development, *Front. Cell Dev.* Biol. 6: 59 doi:10.3389/fcell.2018.00059

Balkwill, F. and Mantovani, A. (2001) Inflammation and cancer: back to Virchow? *The Lancet* 357: 539–545 doi:10.1016/S0140-6736(00)04046-0

Tippimanchai, D.D., Nolan, K., Poczobutt, J. et al. (2018) Adenoviral vectors transduce alveolar macrophages in lung cancer models, *Oncoimmunology* 7: e1438105 doi:10.1080/2162402X.2018.1438105

Henry, C.J., Sedjo, R.L., Rozhok, A. et al. (2015) Lack of significant association between serum inflammatory cytokine profiles and the presence of colorectal adenoma, *BMC Cancer* 15: 123 doi:10.1186/s12885-015-1115-2

Krall, J.A., Reinhardt, F., Mercury, O.A. et al. (2018) The systemic response to surgery triggers the outgrowth of distant immune-controlled tumors in mouse models of dormancy, *Science* Translational Medicine 10: eaan3464 doi:10.1126/scitranslmed.aan3464

Marusyk, A., Casás-Selves, M., Henry, C.J. et al. (2009) Irradiation alters selection for oncogenic mutations in hematopoietic progenitors, *Cancer Research* 69: 7262–7269 doi:10.1158/0008-5472.CAN-09-0604

Risques, R.A. and Kennedy, S.R. (2018) Aging and the rise of somatic cancer-associated mutations in normal tissues, *PLoS Genet* 14: e1007108 doi: 10.1371/journal.pgen.1007108

Bissell, M., Hines, W. (2011) Why don't we get more cancer? A proposed role of the microenvironment in restraining cancer progression, *Nat Med* 17: 320–329 doi:10.1038/nm.2328

Maffini, M.V., Soto, A.M., Calabro, J.M. et al. (2004) The stroma as a crucial target in rat mammary gland carcinogenesis, *Journal of Cell Science* 117: 1495–1502 doi:10.1242/jcs.01000

Rubin, H. (1985) Cancer as a dynamic developmental disorder, *Cancer Res* 45: 2935-2942

Dong, X., Milholland, B. & Vijg, J. (2016) Evidence for a limit to human lifespan, *Nature* 538: 257–259 doi:10.1038/nature19793

Greaves, M. (2018) A causal mechanism for childhood acute lymphoblastic leukemia, *Nat Rev Cancer* 18: 471–484 doi:10.1038/s41568-018-0015-6

Wilson, B.T., Douglas, S.F., and Polvikoski, T. (2010) Astrocytoma in a Breast Cancer Lineage: Part of the BRCA2 Phenotype? *Journal of Clinical Oncology* 28: e596-e598 doi:10.1200/jco.2010.28.9173

Wang, L., Ji, Y., Hu, Y. et al. (2019) The architecture of intra-organism mutation rate variation in plants, *PLoS Biol* 17: e3000191 doi:10.1371/journal.pbio.3000191

Tomasetti, C. and Vogelstein, B. (2015) Variation in cancer risk among tissues can be explained by the number of stem cell divisions, *Science* 347: 78–81 doi: 10.1126/science.1260825

Tomasetti, C., Li, L. and Vogelstein, B. (2017) Stem cell divisions, somatic mutations, cancer etiology, and cancer prevention, *Science* 355: 1330–1334 doi:10.1126/science.aaf9011

Blokzijl, F., de Ligt, J., Jager, M. et al. (2016) Tissue-specific mutation accumulation in human adult stem cells during life, *Nature* 538: 260–264 doi:10.1038/nature19768

Buell, P. (1973) Changing incidence of breast cancer in Japanese-American women, *JNCI: Journal of the National Cancer Institute* 51: 1479–1483 doi:10.1093/jnci/51.5.1479

DCIS Precision website dcisprecision.org

## Chapter 6

Jamieson A. (2010) Scientists hail "penicillin moment" in cancer treatment, *Daily Telegraph* (published online September 15, 2010) bit.ly/39F6FJ1

Ledford, H. (2010) Rare victory in fight against melanoma, *Nature* 467: 140–141 doi:10.1038/467140b

Chamberlain G. (2006) British maternal mortality in the 19th and early 20th centuries, *Journal of the Royal Society of Medicine* 99: 559–563 doi:10.1258/jrsm.99.11.559

Yachida, S., Jones, S., Bozic, I. et al. (2010) Distant metastasis occurs late during the genetic evolution of pancreatic cancer, *Nature* 467: 1114–1117 doi:10.1038/nature09515

Tao, Y., Ruan, J., Yeh, S.H. et al. (2011) Rapid growth of a hepatocellular carcinoma and the driving mutations revealed by cell-population genetic analysis of whole-genome data, *Proceedings of the National Academy of Sciences USA* 108: 12042–12047 doi:10.1073/pnas.1108715108

Campbell, P.J., Pleasance, E.D., Stephens, P.J. et al. (2008) Subclonal phylogenetic structures in cancer revealed by ultra-deep sequencing.

*Proceedings of the National Academy of Sciences USA* 105: 13081–13086 doi:10.1073/pnas.0801523105

Mullighan, C.G., Phillips, L.A., Su, X. et al. (2008) Genomic analysis of the clonal origins of relapsed acute lymphoblastic leukemia, *Science* 322: 1377–1380 doi:10.1126/science.1164266

Inukai, M., Toyooka, S., Ito, S. et al. (2006) Presence of epidermal growth factor receptor gene T790M mutation as a minor clone in non–small cell lung cancer, *Cancer Research* 66: 7854-7858 doi:10.1158/0008-5472. CAN-06-1951

Navin, N., Kendall, J., Troge, J. et al. (2011) Tumor evolution inferred by single-cell sequencing, *Nature* 472: 90–94 doi:10.1038/nature09807

Gerlinger, M., Rowan, A.J., Horswell, S. et al. (2012) Intratumor heterogeneity and branched evolution revealed by multiregion sequencing, *N Engl J Med* 366: 883–892 doi:10.1056/NEJMoa1113205

Darwin, C. R. (1881) *The Formation of Vegetable Mold, Through the Action of Worms*, John Murray, London, Chapter 1, p26

Lu, Y., Wajapeyee, N., Turker, M.S., and Glazer, P.M. (2014) Silencing of the DNA mismatch repair gene MLH1 induced by hypoxic stress in a pathway dependent on the histone demethylase LSD1, *Cell Reports* 8: 501–513 doi:10.1016/j.celrep.2014.06.035

Ding, L., Ley, T., Larson, D. et al. (2012) Clonal evolution in relapsed acute myeloid leukemia revealed by whole genome sequencing, *Nature* 481: 506–510 doi:10.1038/nature10738

Hunter C., Smith, R., Cahill, D.P. et al. (2006) A hypermutation phenotype and somatic MSH6 mutations in recurrent human malignant gliomas after alkylator chemotherapy, *Cancer Res.* 66: 3987–91 doi: 10.1158/0008-5472.CAN-06-0127

Russo, M., Crisafulli, G., Sogari, A. et al. (2019) Adaptive mutability of colorectal cancers in response to targeted therapies, *Science* 366: 1473–1480 doi:10.1126/science.aav4474

Keats, J.J., Chesi, M., Egan, J.B. et al. (2012) Clonal competition with alternating dominance in multiple myeloma, *Blood* 120: 1067–1076 doi:10.1182/blood-2012-01-405985

Morrissy, A. S., Garzia, L., Shih, D. J. et al. (2016) Divergent clonal selection dominates medulloblastoma at recurrence, *Nature* 529: 351–357 doi:10.1038/nature16478

Nowell, P.C. (1976) The clonal evolution of tumor cell populations, *Science* 194: 23–28 doi:10.1126/science.959840

Aktipis, C.A., Kwan, V.S.Y., Johnson, K.A. et al. (2011) Overlooking evolution: a systematic analysis of cancer relapse and therapeutic resistance research, *PLoS ONE* 6: e26100 doi:10.1371/journal.pone.0026100

Smith, M.P. and Harper, D.A.T. (2013) Causes of the Cambrian Explosion, *Science* 341: 1355–1356 doi:10.1126/science.1239450

Notta, F., Chan-Seng-Yue, M., Lemire, M. et al. (2016) A renewed model of pancreatic cancer evolution based on genomic rearrangement patterns, *Nature* 538: 378–382 doi:10.1038/nature19823

Chen, G., Bradford, W.D., Seidel, C.W. and Li, R. (2012) Hsp90 stress potentiates rapid cellular adaptation through induction of aneuploidy, *Nature* 482: 246–250 doi:10.1038/nature10795

Potapova, T. A., Zhu, J. and Li, R. (2013). Aneuploidy and chromosomal instability: a vicious cycle driving cellular evolution and cancer genome chaos, *Cancer Metastasis Reviews* 32: 377–389 doi:10.1007/s10555-013-9436-6

Chen, G., Rubinstein, B. and Li, R. (2012). Whole chromosome aneuploidy: big mutations drive adaptation by phenotypic leap, *BioEssays* 34: 893–900 doi:10.1002/bies.201200069

Baker, D., Jeganathan, K., Cameron, J. et al. (2004) BubR1 insufficiency causes early onset of aging-associated phenotypes and infertility in mice, *Nat Genet* 36: 744–749 doi:10.1038/ng1382

Baker, D.J., Dawlaty, M.M., Wijshake, T. et al. (2013) Increased expression of BubR1 protects against aneuploidy and cancer and extends healthy lifespan, *Nature Cell Biology* 15: 96–102 doi:10.1038/ncb2643

Sackton, K., Dimova, N., Zeng, X. et al. (2014) Synergistic blockade of mitotic exit by two chemical inhibitors of the APC/C, *Nature* 514: 646–649 doi:10.1038/nature13660

Martincorena, I. and Campbell, P.J. (2015) Somatic mutation in cancer and normal cells, *Science* 349: 1483–1489 doi:10.1126/science.aab4082

Stephens, P. J., Greenman, C. D., Fu, B. et al. (2011) Massive genomic rearrangement acquired in a single catastrophic event during cancer development, *Cell* 144: 27–40 doi:10.1016/j.cell.2010.11.055

Wu, S., Turner, K.M., Nguyen, N. et al. (2019) Circular ecDNA promotes accessible chromatin and high oncogene expression, *Nature* 575: 699–703 doi:10.1038/s41586-019-1763-5

Garsed, D.W., Marshall, O.J., Corbin, V.D.A. et al. (2014) The architecture and evolution of cancer neochromosomes, *Cancer Cell* 26: 653-667 doi:10.1016/j.ccell.2014.09.010

Sheltzer, J.M., Ko, J.H., Replogle, J.M. et al. (2017) Single-chromosome gains commonly function as tumor suppressors, *Cancer Cell* 31: 240–255 doi:10.1016/j.ccell.2016.12.004

Relationship between incorrect chromosome number and cancer is reassessed after surprising experiments (2017). Cold Spring Harbor Laboratory website (published online January 12, 2017) bit.ly/2ZZwAXy

Thompson, S.L. and Compton, D.A. (2011) Chromosomes and cancer cells, *Chromosome Research* 19: 433–444 doi:10.1007/s10577-010-9179-y

IJdo, J.W., Baldini, A., Ward, D.C. et al. (1991) Origin of human chromosome 2: an ancestral telomere-telomere fusion, *Proceedings of the National Academy of Sciences USA* 88: 9051-9055 doi:10.1073/pnas.88.20.9051

Van Valen, L.M. and Maiorana, V.C. (1991). HeLa, a new microbial species, *Evolutionary Theory & Review* 10: 71–74

Adey, A., Burton, J., Kitzman, J. et al. (2013) The haplotype-resolved genome and epigenome of the aneuploid HeLa cancer cell line, *Nature* 500: 207–211 doi:10.1038/nature12064

Landry, J.J., Pyl, P.T., Rausch, T. et al. (2013) The genomic and transcriptomic landscape of a HeLa cell line, *G3* 3: 1213–1224 doi:10.1534/g3.113.005777

Nelson-Rees, W.A., Daniels, D.W. and Flandermeyer, R.R. (1981) Cross-contamination of cells in culture, *Science* 212: 446–452 doi:10.1126/science.6451928

Oransky, I. and Marcus, A. (2016) Thousands of studies used the wrong cells, and journals are doing nothing, *STAT* (published online July 21, 2016) bit.ly/39GMNVR

Neimark, J. (2015) Line of attack, *Science* 347: 938-940 doi:10.1126/science.347.6225.938

Masters, J. (2002) HeLa cells 50 years on: the good, the bad and the ugly, *Nat Rev Cancer* 2: 315–319 doi:10.1038/nrc775

Hanahan, D. and Weinberg, R.A. (2000) The hallmarks of cancer, *Cell* 100:57–70 doi:10.1016/s0092-8674(00)81683-9

Hanahan, D. and Weinberg, R. (2011) Hallmarks of cancer: the next generation, *Cell* 144:646–674 doi:10.1016/j.cell.2011.02.013

Freeman, S. (2008) How dictators work, *How Stuff Works* (published online April 2, 2008) bit.ly/2tsgmKn

Wong, K., van der Weyden, L., Schott, C.R. et al. (2019) Cross-species genomic landscape comparison of human mucosal melanoma with canine oral and equine melanoma, *Nature* Communications 10: 353 doi:10.1038/s41467-018-08081-1

Swanton, C. (2015) Cancer evolution constrained by mutation order, *N Engl J Med* 372: 661–663 doi:10.1056/NEJMe1414288

CDC, "More People in the United States Dying from Antibiotic-Resistant Infections than Previously Estimated," https://www.cdc.gov/media/releases/2019/p1113-antibiotic-resistant.html#:~:text=The%20Centers%20for%20Disease%20Control,the%20United%20States%20each%20year.

## Chapter 7

Rosenthal, R., Cadieux, E.L., Salgado, R. et al. (2019) Neoantigen-directed immune escape in lung cancer evolution, *Nature* 567: 479–485 doi:10.1038/s41586-019-1032-7

Coudray, N., Ocampo, P.S., Sakellaropoulos, T. et al. (2018) Classification and mutation prediction from non–small cell lung cancer histopathology images using deep learning, *Nat Med* 24: 1559–1567 doi:10.1038/s41591-018-0177-5

Warburg, O. (1956) On the origin of cancer cells, *Science* 123: 309–314 doi:10.1126/science.123.3191.309

Dvorak, H.F. (1986) Tumors: wounds that do not heal, *N Engl J Med* 315: 1650-1659 doi:10.1056/NEJM198612253152606

Kortlever, R.M., Sodir, N.M., Wilson, C.H. et al. (2017) Myc cooperates with ras by programming inflammation and immune suppression, *Cell* 171: 1301–1315.e14 doi:10.1016/j.cell.2017.11.013

Sambon, L. W. (1924) The elucidation of cancer, *Proceedings of the Royal Society of Medicine* 17: 77–124 doi:10.1177/003591572401701607

Folkman, J. (1971) Tumor angiogenesis: therapeutic implications, *N Engl J Med* 285: 1182–1186 doi:10.1056/NEJM197111182852108

Folkman, J., Merler, E., Abernathy, C. and Williams, G. (1971) Isolation of a tumor factor responsible for angiogenesis, *The Journal of Experimental Medicine* 133: 275–288 doi:10.1084/jem.133.2.275

Kolata, G. (1998) HOPE IN THE LAB: A special report. A cautious awe greets drugs that eradicate tumors in mice, *New York Times* (published May 3, 1998) nyti.ms/36p1FWQ

Maniotis, A. J., Folberg, R., Hess, A. et al. (1999) Vascular channel formation by human melanoma cells in vivo and in vitro: vasculogenic mimicry, *The American Journal of Pathology* 155: 739–752 doi:10.1016/S0002-9440(10)65173-5

Wagenblast, E., Soto, M., Gutiérrez-Ángel, S. et al. (2015) A model of breast cancer heterogeneity reveals vascular mimicry as a driver of metastasis, *Nature* 520: 358–362 doi:10.1038/nature14403

Cleary, A.S., Leonard, T.L., Gestl, S.A. and Gunther, E.J. (2014) Tumor cell heterogeneity maintained by cooperating subclones in Wnt-driven mammary cancers, *Nature* 508: 113–117 doi:10.1038/nature13187

Marusyk, A., Tabassum, D., Altrock, P. et al. (2014) Non-cell-autonomous driving of tumor growth supports sub-clonal heterogeneity, *Nature* 514: 54–58 doi:10.1038/nature13556

Laelaps (2015) When monkeys surfed to South America, *National Geographic* (published online February 5, 2015) on.natgeo.com/2SVckVe

Bond, M., Tejedor, M., Campbell, K. et al. (2015) Eocene primates of South America and the African origins of New World monkeys, *Nature* 520: 538–541 doi:10.1038/nature14120

Freeman, M. D., Gopman, J. M., & Salzberg, C. A. (2018) The evolution of mastectomy surgical technique: from mutilation to medicine, *Gland Surgery* 7: 308–315 doi:10.21037/gs.2017.09.07

Fidler I.J. and Poste, G. (2008) The "seed and soil" hypothesis revisited, *The Lancet Oncology* 9: 808 doi: 10.1016/S1470-2045(08)70201-8

Reinshagen, C., Bhere, D., Choi, S.H. et al.(2018) CRISPR-enhanced engineering of therapy-sensitive cancer cells for self-targeting of primary and metastatic tumors, *Science Translational Medicine* 10: eaao3240 doi:10.1126/scitranslmed.aao3240

Peinado, H., Zhang, H., Matei, I. et al. (2017) Pre-metastatic niches: organ-specific homes for metastases, *Nat Rev Cancer* 17: 302–317 doi:10.1038/nrc.2017.6

Kaplan, R. N., Riba, R. D., Zacharoulis, S. et al. (2005) VEGFR1-positive haematopoietic bone marrow progenitors initiate the pre-metastatic niche, *Nature* 438: 820–827 doi:10.1038/nature04186

Albrengues, J., Shields, M. A., Ng, D. et al. (2018) Neutrophil extracellular traps produced during inflammation awaken dormant cancer cells in mice, *Science* 361: eaao4227 doi:10.1126/science.aao4227

Sanz-Moreno, V. and Balkwill, F.R. (2009) Mets and NETs: the awakening force, *Immunity* 49: 798-800 doi:10.1016/j.immuni.2018.11.009

Ridker, P.M., Everett, B.M., Thuren, T. et al. (2017) Anti-inflammatory therapy with canakinumab for atherosclerotic disease, *N Engl J Med* 377: 1119-1131 doi:10.1056/NEJMoa1707914

Oswald, L., Grosser, S., Smith, D. M. and Käs, J. A. (2017) Jamming transitions in cancer, *Journal of Physics D* 50: 483001 doi:10.1088/1361-6463/aa8e83

Fojo, T. (2018) Desperation oncology, *Seminars in Oncology* 45: 105–106 doi:10.1053/j.seminoncol.2018.08.001

Kaiser, J. (2019) New drugs that unleash the immune system on cancers may backfire, fueling tumor growth, *Science* (published online March 28, 2019) doi:10.1126/science.aax5021

Champiat, S., Dercle, L., Ammari, S. et al.(2017) Hyperprogressive disease is a new pattern of progression in cancer patients treated by anti-PD-1/PD-L1, *Clin Cancer Res* 23: 1920–1928 doi:10.1158/1078-0432.CCR-16-1741

Obradović, M.M.S., Hamelin, B., Manevski, N. et al. (2019) Glucocorticoids promote breast cancer metastasis, *Nature* 567: 540–544 doi:10.1038/s41586-019-1019-4

Greaves, M. (2018) A causal mechanism for childhood acute lymphoblastic leukemia, *Nat Rev Cancer* 18: 471–484 doi:10.1038/s41568-018-0015-6

Gopalakrishnan, V., Helmink, B. A., Spencer, C. N. et al. (2018). The influence of the gut microbiome on cancer, immunity, and cancer immunotherapy, *Cancer Cell* 33: 570–580 doi:10.1016/j.ccell.2018.03.015

Alexander, J., Wilson, I., Teare, J. et al. (2017) Gut microbiota modulation of chemotherapy efficacy and toxicity, *Nat Rev Gastroenterol Hepatol* 14: 356–365 doi:10.1038/nrgastro.2017.20

Richards, S.E. (2019) How the microbiome could be the key to new cancer treatments, *Smithsonian Magazine* (published online March 8, 2019) bit.ly/37GFLii

Gharaibeh, R.Z. and Jobin, C. (2019) Microbiota and cancer immunotherapy: in search of microbial signals, *Gut* 68:385–388 doi:10.1136/gutjnl-2018-317220

Zheng, Y., Wang, T., Tu, X. et al. (2019) Gut microbiome affects the response to anti-PD-1 immunotherapy in patients with hepatocellular carcinoma, *J. Immunotherapy Cancer* 7: 193 doi:10.1186/s40425-019-0650-9

Dambuza, I.M. and Brown, G.D. (2019) Fungi accelerate pancreatic cancer, *Nature* 574: 184–185 doi:10.1038/d41586-019-02892-y

Aykut, B., Pushalkar, S., Chen, R. et al. (2019) The fungal mycobiome promotes pancreatic oncogenesis via activation of MBL, *Nature* 574: 264–267 doi:10.1038/s41586-019-1608-2

Saus, E. Iraola-Guzmán, S., Willis, J.R. et al. (2019) Microbiome and colorectal cancer: Roles in carcinogenesis and clinical potential, *Molecular Aspects of Medicine* 69: 93-106 doi:10.1016/j.mam.2019.05.001

Rubinstein, M.R., Baik, J.E., Lagana, S.M. et al. (2019) Fusobacterium nucleatum promotes colorectal cancer by inducing Wnt/ catenin modulator Annexin A1, *EMBO Rep* 20: e47638 doi:10.15252/embr.201847638

Orritt, R. (2016) Why has science seemingly changed its mind on night shifts and breast cancer? Cancer Research UK Science blog (published online October 14, 2016) bit.ly/2umMUpx

Yang, Y., Adebali, O., Wu, G. et al. (2018) Cisplatin-DNA adduct repair of transcribed genes is controlled by two circadian programs in mouse tissues, *Proceedings of the National Academy of Sciences USA* 115: E4777-E4785 doi:10.1073/pnas.1804493115

Guevara-Aguirre, J., Balasubramanian, P., Guevara-Aguirre, M. et al. (2011) Growth hormone receptor deficiency is associated with a major reduction in pro-aging signaling, cancer, and diabetes in humans, *Science Translational Medicine* 70: 70ra13 doi:10.1126/scitranslmed.3001845

Bowes, P. (2016) The experimental diet that mimics a rare genetic mutation, *Mosaic* (published online April 11 2016) bit.ly/2QODuuh

Cornaro, A. translated by Fudemoto, H. (2014) *Writings on the Sober Life: The Art and Grace of Living Long*, University of Toronto Press, p22

## Chapter 8

Noveski, P., Madjunkova, S., Sukarova Stefanovska, E. et al. (2016). Loss of Y chromosome in peripheral blood of colorectal and prostate cancer patients, *PloS ONE* 11: e0146264 doi:10.1371/journal.pone.0146264

Dumanski, J.P., Rasi, C., Lönn, M. et al. (2015) Smoking is associated with mosaic loss of chromosome Y, *Science* 347: 81–83 doi:10.1126/science.1262092

Yang, W., Warrington, N.M., Taylor, S.J. et al. (2019) Sex differences in GBM revealed by analysis of patient imaging, transcriptome, and survival data, *Science Translational Medicine* 11: eaao5253 doi:10.1126/scitranslmed.aao5253

Venkatesh, H., Morishita, W., Geraghty, A. et al. (2018) Excitatory synapses between presynaptic neurons and postsynaptic glioma cells promote glioma progression, *Neuro-Oncology* 20: vi257–vi258 doi:10.1093/neuonc/noy148.1069

Gillespie, S. and Monje, M. (2018) An active role for neurons in glioma progression: making sense of Scherer's structures, *Neuro-Oncology* 20: 1292–1299 doi:10.1093/neuonc/noy083

Gast, C.E., Silk, A.D., Zarour, L. et al. (2018) Cell fusion potentiates tumor heterogeneity and reveals circulating hybrid cells that correlate with stage and survival, *Science Advances* 4: eaat7828 doi:10.1126/sciadv.aat7828

Carter A. (2008) Cell fusion theory: can it explain what triggers metastasis? *J Natl Cancer Inst.* 100: 1279–81 doi:10.1093/jnci/djn336

Lin, K., Torga, G., Sun, Y. et al. (2019) The role of heterogeneous environment and docetaxel gradient in the emergence of polyploid, mesenchymal and resistant prostate cancer cells, *Clin Exp Metastasis* 36: 97–108 doi:10.1007/s10585-019-09958-1

Lu, X. and Kang, Y. (2009) Cell fusion as a hidden force in tumor progression, *Cancer Research* 69: 8536–8539 doi:10.1158/0008-5472.CAN-09-2159

Moore, A. (2012), Cancer: Escape route from a "doomed" host? *Bioessays* 34: 2-2 doi:10.1002/bies.201190072

Clarification of Cancer-Cell Transmission in Tasmania Devil Facial Tumor Disease (2012). Prince Hitachi Prize for Comparative Oncology website bit.ly/2FoF9Bu

Pearse, A., Swift, K. (2006) Transmission of devil facial-tumor disease, *Nature* 439: 549 doi:10.1038/439549a

Siddle, H.V., Kreiss, A., Eldridge, M.D. et al. (2007) Transmission of a fatal clonal tumor by biting occurs due to depleted MHC diversity in a threatened carnivorous marsupial, *Proceedings of the National Academy of Sciences USA* 104: 16221–16226 doi:10.1073/pnas.0704580104

Murchison, E.P., Tovar, C., Hsu, A. et al. (2010) The Tasmanian devil transcriptome reveals Schwann cell origins of a clonally transmissible cancer, *Science* 327: 84–87 doi:10.1126/science.1180616

Murchison, E.P., Schulz-Trieglaff, O.B., Ning, Z. et al. (2012) Genome sequencing and analysis of the Tasmanian devil and its transmissible cancer, *Cell* 148: 780–791 doi:10.1016/j.cell.2011.11.065

Pye, R.J., Pemberton, D., Tovar, C. et al. (2016) A second transmissible cancer in Tasmanian devils, *Proceedings of the National Academy of Sciences USA* 113: 374–379 doi:10.1073/pnas.1519691113

Caldwell, A., Coleby, R., Tovar, C. et al. (2018) The newly-arisen Devil facial tumor disease 2 (DFT2) reveals a mechanism for the emergence of a contagious cancer, *eLife* 7: e35314 doi:10.7554/eLife.35314

Timmins, B. (2019) Tasmanian devils "adapting to coexist with cancer," *BBC News Online* (published online March 30, 2019) bbc.in/39GZsbl

Wells, K., Hamede, R.K., Jones, M.E. (2019) Individual and temporal variation in pathogen load predicts long-term impacts of an emerging infectious disease, *Ecology* 100: e02613 doi:10.1002/ecy.2613

Karlson, A.G. and Mann, F.C. (1952) The transmissible venereal tumor of dogs: observations on forty generations of experimental transfers, *Ann N Y Acad Sci.* 54: 1197–213 doi:10.1111/j.1749-6632.1952.tb39989.x

Das, U. & Das, A.K. (2000) Review of canine transmissible venereal sarcoma, *Vet Res Commun* 24: 545 doi:10.1023/A:1006491918910

Murgia, C., Pritchard, J. K., Kim, S. Y. et al. (2006) Clonal origin and evolution of a transmissible cancer, *Cell* 126: 477–487 doi:10.1016/j.cell.2006.05.051

Murchison, E.P., Wedge, D.C., Alexandrov, L.B. et al. (2014) Transmissible dog cancer genome reveals the origin and history of an ancient cell lineage, *Science* 343: 437–440 doi:10.1126/science.1247167

Parker, H.G., & Ostrander, E.A. (2014) Hiding in plain view—an ancient dog in the modern world, *Science* 343: 376–378 doi:10.1126/science.1248812

Cranage, A. (2018) Chernobyl: Chasing a "catching" cancer. Wellcome Sanger Institute blog (published online December 7, 2018) bit.ly/2T5sg7N

Metzger, M. J., Reinisch, C., Sherry, J. and Goff, S. P. (2015) Horizontal transmission of clonal cancer cells causes leukemia in soft-shell clams, *Cell* 161: 255–263 doi:10.1016/j.cell.2015.02.042

Metzger, M., Villalba, A., Carballal, M. et al. (2016) Widespread transmission of independent cancer lineages within multiple bivalve species, *Nature* 534: 705–709 doi:10.1038/nature18599

Yonemitsu, M.A., Giersch, R.M., Polo-Prieto, M. et al. (2019) A single clonal lineage of transmissible cancer identified in two marine mussel species in South America and Europe, *eLife* 8: e47788 doi:10.7554/eLife.47788

Greaves, M.F., Maia, A.T., Wiemels, J.L. and Ford, A.M. (2003) Leukemia in twins: lessons in natural history, *Blood* 102: 2321–2333 doi:10.1182/blood-2002-12-3817

Greaves, M. and Hughes, W. (2018) Cancer cell transmission via the placenta, *Evolution, Medicine, and Public Health* 1: 106–115 doi:10.1093/emph/eoy011

Desai, R., Collett, D., Watson, C.J.E. et al. (2014) Estimated risk of cancer transmission from organ donor to graft recipient in a national transplantation registry, *Br J Surg* 101: 768-774 doi:10.1002/bjs.9460

Matser, YAH, Terpstra, ML, Nadalin, S, et al. (2018) Transmission of breast cancer by a single multiorgan donor to 4 transplant recipients, *Am J Transplant* 18: 1810–1814 doi:10.1111/ajt.14766

Gärtner, H-V., Seidl, C., Luckenbach, C. et al. (1996) Genetic analysis of a sarcoma accidentally transplanted from a patient to a surgeon, *N Engl J Med* 335: 1494–1497 doi:10.1056/NEJM199611143352004

Gugel, E.A. and Sanders, M.E. (1986) Needle-stick transmission of human colonic adenocarcinoma, *N Engl J Med* 315: 1487 doi:10.1056/NEJM198612043152314

Hornblum, A.M. (2013) NYC's forgotten cancer scandal, *New York Post* (published online December 28, 2013) bit.ly/2SSOp8X

Hornblum, A.M. (1997) They were cheap and available: prisoners as research subjects in twentieth century America, *BMJ* 315: 1437 doi:10.1136/bmj.315.7120.1437

Southam, C.M. and Moore, A.E. (1958) Induced immunity to cancer cell

homografts in man, *Annals of the New York Academy of Sciences* 73: 635–653 doi:10.1111/j.1749-6632.1959.tb40840.x
Osmundsen, J.A. (1964) Many scientific experts condemn ethics of cancer injection, *New York Times* (published January 26, 1964) nyti.ms/2MYhaxo
Scanlon, E.F., Hawkins, R.A., Fox, W.W. and Smith, W.S. (1965) Fatal homotransplanted melanoma, *Cancer* 18:782–9 doi:10.1002/1097-0142
Muehlenbachs, A., Bhatnagar, J., Agudelo, C.A. et al. (2015) Malignant transformation of *Hymenolepis nana* in a human host, *N Engl J Med* 373: 1845-1852 doi:10.1056/NEJMoa1505892
Fabrizio, A.M. (1965) An induced transmissible sarcoma in hamsters: eleven-year observation through 288 passages, *Cancer Research* 25: 107–117
Banfield, W.G., Woke, P.A., Mackay, C.M., and Cooper, H.L. (1965) Mosquito transmission of a reticulum cell sarcoma of hamsters, *Science* 148: 1239–1240 doi:10.1126/science.148.3674.1239

## Chapter 9

Marquart, J., Chen, E.Y. and Prasad V. (2018) Estimation of the percentage of US patients with cancer who benefit from genome-driven oncology, *JAMA Oncol.* 4: 1093–1098 doi:10.1001/jamaoncol.2018.1660
Abola, M.V., Prasad, V. (2016) The use of superlatives in cancer research, *JAMA Oncol.* 2: 139–141 doi:10.1001/jamaoncol.2015.3931
Kuderer, N. M., Burton, K. A., Blau, S. et al. (2017) Comparison of 2 commercially available next-generation sequencing platforms in oncology, *JAMA Oncology* 3: 996–998 doi:10.1001/jamaoncol.2016.4983
Prahallad, A., Sun, C., Huang, S. et al. (2012) Unresponsiveness of colon cancer to BRAF(V600E) inhibition through feedback activation of EGFR, *Nature* 483: 100–103 doi:10.1038/nature10868
Prasad V. (2017) Overestimating the benefit of cancer drugs, *JAMA Oncol.* 3: 1737–1738 doi: 10.1001/jamaoncol.2017.0107
Salas-Vega, S., Iliopoulos, O. and Mossialos, E. (2017) Assessment of overall survival, quality of life, and safety benefits associated with new cancer medicines, *JAMA Oncol.* 3: 382–390 doi:10.1001/jamaoncol.2016.4166
Fojo, T., Mailankody, S. and Lo, A. (2014) Unintended consequences of expensive cancer therapeutics—the pursuit of marginal indications and a Me-Too mentality that stifles innovation and creativity: The John Conley Lecture, *JAMA Otolaryngol Head Neck Surg.* 140:1225–1236 doi:10.1001/jamaoto.2014.1570
Lomangino, K. (2017) "Not statistically significant but clinically meaningful": A researcher calls "BS" on cancer drug spin, *Health News Review* (published online March 24, 2017) bit.ly/35riDTq
Oyedele, A. (2014) 19 of the Most Expensive Substances In the World, *Business Insider* (published online September 22, 2014) bit.ly/36kqx1W
Kim, C. and Prasad, V. (2015) Cancer drugs approved on the basis of a

surrogate end point and subsequent overall survival: an analysis of 5 years of US Food and Drug Administration approvals, *JAMA Intern Med.* 175: 1992–4 doi:10.1001/jamainternmed.2015.5868

Prasad, V., McCabe, C. and Mailankody, S. (2018) Low-value approvals and high prices might incentivize ineffective drug development, *Nat Rev Clin Oncol* 15: 399–400 doi:10.1038/s41571-018-0030-2

Prasad, V. and Mailankody, S. (2017) Research and development spending to bring a single cancer drug to market and revenues after approval, *JAMA Intern Med.* 177: 1569–1575 doi:10.1001/jamainternmed.2017.3601

Prasad, V. (2016) Perspective: The precision-oncology illusion, *Nature* 537: S63 doi:10.1038/537S63a

Perelson, A.S., Neumann, A.U., Markowitz, M. et al (1996) HIV-1 dynamics in vivo: virion clearance rate, infected cell life-span, and viral generation time, *Science* 271: 1582–6 doi:10.1126/science.271.5255.1582

The Antiretroviral Therapy Cohort Collaboration (2017) Survival of HIV-positive patients starting antiretroviral therapy between 1996 and 2013: a collaborative analysis of cohort studies, *The Lancet HIV* 4: PE349-E356 doi:10.1016/S2352-3018(17)30066-8

Clarke, P.A., Roe, T., Swabey, K. et al. (2019) Dissecting mechanisms of resistance to targeted drug combination therapy in human colorectal cancer, *Oncogene* 38: 5076–5090 doi:10.1038/s41388-019-0780-z

Behan, F.M., Iorio, F., Picco, G. et al. (2019) Prioritization of cancer therapeutic targets using CRISPR–Cas9 screens, *Nature* 568: 511–516 doi:10.1038/s41586-019-1103-9

Momen, S., Fassihi, H., Davies, H.R. et al. (2019) Dramatic response of metastatic cutaneous angiosarcoma to an immune checkpoint inhibitor in a patient with xeroderma pigmentosum: whole-genome sequencing aids treatment decision in end-stage disease, *Cold Spring Harb Mol Case Stud* 5: a004408 doi:10.1101/mcs.a004408

Ritchie, Hannah, Cancer death rates are falling; five-year survival rates are rising, https://ourworldindata.org/cancer-death-rates-are-falling-five-year-survival-rates-are-rising

## Chapter 10

Markus, C. and McFeely, S. (2018) *Avengers: Infinity War*, dir. Russo, A. and Russo, J. Marvel Studios

Enriquez-Navas, P.M., Wojtkowiak, J.W. and Gatenby, R.A. (2015) Application of evolutionary principles to cancer therapy, *Cancer Res.* 75: 4675–80 doi:10.1158/0008-5472.CAN-15-1337

Enriquez-Navas, P.M., Kam, Y., Das, T. et al. (2016) Exploiting evolutionary principles to prolong tumor control in preclinical models of breast cancer, *Science Translational Medicine* 8: 327ra24 doi:10.1126/scitranslmed.aad7842

Wang, L. & Bernards, R. (2018) Taking advantage of drug resistance, a

new approach in the war on cancer, *Front. Med.* 12: 490 doi:10.1007/s11684-018-0647-7
Gatenby, R.A., Silva, A.S., Gillies, R.J. and Frieden, B.R. (2009) Adaptive therapy, *Cancer Res.* 69: 4894–903 doi:10.1158/0008-5472.CAN-08-3658
Zhang, J., Cunningham, J.J., Brown, J.S., and Gatenby, R.A. (2017) Integrating evolutionary dynamics into treatment of metastatic castrate-resistant prostate cancer, *Nat Commun.* 8: 1816 doi:10.1038/s41467-017-01968-5
Khan, K.H., Cunningham, D., Werner, B. et al. (2018) Longitudinal liquid biopsy and mathematical modeling of clonal evolution forecast time to treatment failure in the PROSPECT-C Phase II colorectal cancer clinical trial, *Cancer Discov.* 8:1270-1285 doi:10.1158/2159-8290.CD-17-0891
Luo, H., Zhao, Q., Wei, W. et al (2020) Circulating tumor DNA methylation profiles enable early diagnosis, prognosis prediction, and screening for colorectal cancer, *Science Translational Medicine* 12: eaax7533 doi: 10.1126/scitranslmed.aax7533
Kam, Y., Das, T., Tian, H. et al. (2015) Sweat but no gain: inhibiting proliferation of multidrug resistant cancer cells with "ersatzdroges," *International Journal of Cancer* 136: E188–E196 doi:10.1002/ijc.29158
Merlo, L.M.F., Pepper, J.W., Reid, B.J. and Maley, C.C. (2006) Cancer as an evolutionary and ecological process, *Nat. Rev. Cancer* 6: 924–935 doi:10.1038/nrc2013
Gatenby, R.A., Brown, J. and Vincent, T. (2009) Lessons from applied ecology: cancer control using an evolutionary double bind, *Cancer Res* 69: 7499–7502 doi:10.1158/0008-5472.CAN-09-1354
Merlo, L.M., Kosoff, R.E., Gardiner, K.L. and Maley C.C. (2011) An in vitro co-culture model of esophageal cells identifies ascorbic acid as a modulator of cell competition. *BMC Cancer* 11: 461 doi:10.1186/1471-2407-11-461
Maley, C.C., Reid, B.J. and Forrest S. (2004) Cancer prevention strategies that address the evolutionary dynamics of neoplastic cells: simulating benign cell boosters and selection for chemosensitivity, *Cancer Epidemiol Biomarkers Prev* 13: 1375–84
Gatenby, R. and Brown, J.S. (2019) Eradicating metastatic cancer and the evolutionary dynamics of extinction, *Preprints* doi:10.20944/preprints201902.0011.v1
Gatenby, R.A., Artzy-Randrup, Y., Epstein, T. et al. (2019) Eradicating metastatic cancer and the eco-evolutionary dynamics of Anthropocene extinctions, *Cancer Research* doi:10.1158/0008-5472.CAN-19-1941
Heisman, R. (2016) The sad story of Booming Ben, last of the heath hens, *JSTOR Daily* (published online March 2, 2016) bit.ly/35mKZhx
Sta ková, K., Brown, J.S., Dalton, W.S. and Gatenby, R.A. (2019) Optimizing cancer treatment using game theory: a review, *JAMA Oncol* 5:96–103 doi:10.1001/jamaoncol.2018.3395
Rosenheim J. A. (2018). Short- and long-term evolution in our arms race with

cancer: why the war on cancer is winnable, *Evolutionary Applications* 11(6), 845–852 doi:10.1111/eva.12612
Repurposing Drugs in Oncology (Re-DO) redoproject.org

## Chapter 11

Baker, S. G., Cappuccio, A., & Potter, J. D. (2010). Research on early-stage carcinogenesis: are we approaching paradigm instability? *Journal of Clinical Oncology* 28: 3215–3218 doi:10.1200/JCO.2010.28.5460

Maley, C., Aktipis, A., Graham, T. et al. (2017) Classifying the evolutionary and ecological features of neoplasms, *Nat Rev Cancer* 17: 605–619 doi:10.1038/nrc.2017.69

Helmneh, M. Sineshaw, H.M, Jemal, A., Ng, K. et al. (2019) Treatment patterns among de novo metastatic cancer patients who died within 1 month of diagnosis, *JNCI Cancer Spectrum* 3: pkz021 doi:10.1093/jncics/pkz021

Ambroggi, M., Biasini, C., Toscani, I. et al. (2018). Can early palliative care with anticancer treatment improve overall survival and patient-related outcomes in advanced lung cancer patients? A review of the literature, *Supportive Care in Cancer* 26: 2945–2953 doi:10.1007/s00520-018-4184-3

Weeks, J.C., Catalano, P.J., Cronin, A. et al (2012) Patients' expectations about effects of chemotherapy for advanced cancer, *N Engl J Med 367*: 1616–1625 doi:10.1056/NEJMoa1204410

Dobzhansky, T. (1973) Nothing in biology makes sense except in the light of evolution, *The American Biology Teacher* 35: 125–129 doi:10.2307/4444260

Berenblum, I. (1974) Carcinogenesis as a biological problem. *Frontiers of Biology,* 34, Chapter 5.6, p317

McCarthy, M. (2006) New science inspires FDA commissioner Andrew von Eschenbach, *The Lancet* 367: 1649 doi:10.1016/S0140-6736(06)68718-7

Fight On, *The Times* (published online August 30, 2014) bit.ly/37tgEiw

# INDEX

## C

## D

# ABOUT THE AUTHOR

**DR. KAT ARNEY** is a leading British science writer, broadcaster, and public speaker. Her previous books are *Herding Hemingway's Cats: Understanding How Our Genes Work* and *How to Code a Human*. Kat has a degree in natural sciences and a PhD in developmental genetics from Cambridge University and was a key part of the science communications team at Cancer Research UK from 2004 to 2016, cofounding the charity's award-winning science blog and acting as a principal media spokesperson. As a writer, her work has featured in Wired, Daily Mail, Nature, Mosaic, New Scientist, and more. She has fronted several BBC Radio 4 science documentaries, as well as the factual comedy series *Did the Victorians Ruin the World?* and presents the fortnightly Genetics Unzipped podcast.